**EMS Series of Lectures in Mathematics**
Edited by Andrew Ranicki (University of Edinburgh, U.K.)

*EMS Series of Lectures in Mathematics* is a book series aimed at students, professional mathematicians and scientists. It publishes polished notes arising from seminars or lecture series in all fields of pure and applied mathematics, including the reissue of classic texts of continuing interest. The individual volumes are intended to give a rapid and accessible introduction into their particular subject, guiding the audience to topics of current research and the more advanced and specialized literature.

Previously published in this series:
Katrin Wehrheim, *Uhlenbeck Compactness*
Torsten Ekedahl, *One Semester of Elliptic Curves*

Sergey V. Matveev

# Lectures on Algebraic Topology

Translated by Ekaterina Pervova

European Mathematical Society

Author:

Sergey V. Matveev
Department of Computer Topology and Algebra
Chelyabinsk State University
ul. Br. Kashirinykh 129
454021 Chelyabinsk
Russia

Translator:

Ekaterina Pervova
Department of Computer Topology and Algebra
Chelyabinsk State University
ul. Br. Kashirinykh 129
454021 Chelyabinsk
Russia

Originally published by Institute of Computer Science, Moscow-Izhevsk 2003
under the title Лекции по алгебраической топологии

2000 Mathematics Subject Classification: 55-01

ISBN 978-3-03719-023-4

The Swiss National Library lists this publication in The Swiss Book, the Swiss national bibliography, and the detailed bibliographic data are available on the Internet at http://www.helveticat.ch.



Contact address:
European Mathematical Society Publishing House
Seminar for Applied Mathematics
ETH-Zentrum SEW A27
CH-8092 Zürich
Switzerland

Phone: +41 (0)44 632 34 36
Email: info@ems-ph.org
Homepage: www.ems-ph.org

Typeset using the author's TEX files: I. Zimmermann, Freiburg
Printed in Germany

# Preface

Algebraic topology is the study of geometric objects via algebraic methods. Familiarity with its main ideas and methods is quite useful for all undergraduate and graduate students who specialize in any of the many branches of mathematics and physics that have connections to topology, differential geometry, algebra, mathematical analysis, or differential equations. In selecting the content of this book and in writing it the author aspired to reach the following goals:

- to cover those ideas and results that form the backbone of algebraic topology and are sufficient to provide a beautiful, intuitively clear, and logically complete exposition;
- to make the book self-contained, while keeping it reasonably short;
- to make the exposition logically coherent, well-illustrated, and mathematically rigorous, at the same time preserving all the advantages of an informal and lively presentation;
- to structure the text and supplement it with exercises and solutions in such a way that the book becomes a ready-to-use tool for both teachers and students of the subject, as well as a convenient instrument for independent study.

A special attention was devoted to providing explicit algorithms for calculating the homology groups and for manipulating fundamental groups. These subjects are often missing from other books on algebraic topology.

The present book is a revised and slightly extended version of the Russian original publication.

# Contents

# 1

# Elements of homology theory

## 1.1 Categories and functors

One of the main parts of algebraic topology is homology theory, which is a functor from the category of topological spaces to the category of sequences of Abelian groups. Therefore we begin by introducing the notions of a category and a functor.

In order to define a category, we proceed as follows:

1. We specify a certain class of *objects*. Objects may be of any nature.

2. For every ordered pair $A$, $B$ of objects we specify a set of *morphisms* $[A, B]$ of object $A$ to object $B$.

3. For every ordered triple $A, B, C$ of objects we indicate a rule assigning to each pair of morphisms $f \in [A, B]$, $g \in [B, C]$ a third morphism, which belongs to $[A, C]$, is called a *composition* of morphisms $f$, $g$, and is denoted by $gf$. In other words, we define a *composition map* $[A, B] \times [B, C] \to [A, C]$.

**Definition.** The class of objects, the sets of morphisms, and the composition maps thus specified form a *category* if the following axioms hold:

I. Composition of morphisms must be associative, *i.e.* for all triples of morphisms $f \in [A, B]$, $g \in [B, C]$, $h \in [C, D]$ we must have an equality $(hg)f = h(gf)$.

II. For any object $B$ there must be a morphism $\mathrm{Id}_B \in [B, B]$ such that for any two morphisms $f \in [A, B]$, $g \in [B, C]$ the following equalities hold: $\mathrm{Id}_B\, f = f$ and $g\, \mathrm{Id}_B = g$.

Such situations (classes of objects related by morphisms satisfying axioms I, II; *i.e.* categories) arise naturally in many different areas of mathematics. Already at this general level it is possible to give definitions and prove meaningful theorems, which, by virtue of their generality, enjoy a remarkably wide applicability. We restrict ourselves to a very brief introduction into category theory. A more detailed exposition can be found, for instance, in [2].

**Examples of categories**

1. The category of all sets and their maps. The objects of this category are all sets, the morphisms are all possible maps between them.

2. The category of groups and homomorphisms. The objects are groups, and the morphisms are their homomorphisms.

3. The category of Abelian groups and their homomorphisms. The objects of this category are Abelian groups, the morphisms are homomorphisms between them.

It is easy to find other examples of a similar kind: the category of finitely generated groups, the category of rings, and others.

4. The category of topological spaces and continuous maps. The objects are all topological spaces, and the morphisms are their continuous maps.

In all of the above examples the objects are sets, perhaps with additional structures, and the morphisms are maps of sets. However, there exist categories of other types.

5. The category of topological spaces and classes of homotopic maps. The objects of this category are topological spaces, the morphisms are classes of homotopic maps (see definition on page 15). Notice that in this category morphisms are not maps themselves but rather classes of homotopic maps.

Although morphisms do not have to be actual maps, it is quite convenient to denote them in the same way as maps: instead of $f \in [A, B]$, we write $f\colon A \to B$.

**Definition.** Objects $X, Y$ of a category $\mathsf{G}$ are called *isomorphic* if there exist morphisms $f\colon X \to Y$ and $g\colon Y \to X$ such that $fg = \mathrm{Id}_Y$ and $gf = \mathrm{Id}_X$, where $\mathrm{Id}_X$, $\mathrm{Id}_Y$ are the identity morphisms of objects $X$ and $Y$. The morphisms $f, g$ are called *isomorphisms.*

**Example.** Which sets are isomorphic in the category of all sets? It is easy to see that those are exactly sets of equal cardinality, since an isomorphism in this category is nothing other than a bijection.

**Example.** An isomorphism in the category of topological spaces is any homeomorphism; in the category of groups, a group isomorphism. An isomorphism in the category of topological spaces and homotopic maps is called a *homotopy equivalence*. It is also worthwhile to mention an isomorphism in the category of smooth manifolds and smooth maps. That is a *diffeomorphism.*

The usefulness of the notion of a category can be seen already from these examples: a single definition stated in terms of category theory, can replace many corresponding definitions in specific categories. A similar fact is true for theorems as well: if a theorem is proven in categorical terms, then it automatically holds in all specific categories. This observation yields a promising method of obtaining new results.

Let $\mathsf{G}_1$, $\mathsf{G}_2$ be two categories. Suppose that to each object $X$ of the former category we assign an object of the latter category. Let us denote it by $F(X)$. Assume also that to each morphism $f\colon X \to Y$ of $\mathsf{G}_1$ we assign a morphism $f_*\colon F(X) \to F(Y)$ of $\mathsf{G}_2$. Such an assignment is called a *covariant functor* from the category $\mathsf{G}_1$ to the category $\mathsf{G}_2$ if the following axioms hold:

1. If $f$ is an identity morphism, then $f_*$ is also an identity morphism.
2. If a composition $fg$ is well defined, then $(fg)_* = f_* g_*$.

A *contravariant functor* $F$ from a category $\mathsf{G}_1$ to a category $\mathsf{G}_2$ differs from a covariant one in that to each morphism $f\colon X \to Y$ of the category $\mathsf{G}_1$ we assign a morphism $f^*\colon F(Y) \to F(X)$ of the category $\mathsf{G}_2$, *i.e.* one that acts in the opposite direction. The axioms are of course changed in the natural way. In particular, the equality $(fg)_* = f_* g_*$ is replaced by the equality $(fg)^* = g^* f^*$.

**Exercise 1.** Give examples of covariant and contravariant functors.

**Theorem 1.** *Let $F\colon \mathsf{G}_1 \to \mathsf{G}_2$ be a functor from a category $\mathsf{G}_1$ to a category $\mathsf{G}_2$. Suppose that two objects $X$, $Y$ of the category $\mathsf{G}_1$ are isomorphic. Then the objects $F(X)$, $F(Y)$ of the category $\mathsf{G}_2$ are also isomorphic. Equivalently, if the objects $F(X)$, $F(Y)$ of the category $\mathsf{G}_2$ are not isomorphic then neither are the objects $X$, $Y$.*

*Proof.* We limit ourselves to considering a covariant functor. Let $f\colon X \to Y$ and $g\colon Y \to X$ be some isomorphisms between $X$ and $Y$ such that $fg = \mathrm{Id}_Y$ and $gf = \mathrm{Id}_X$. Then it follows immediately from the definition of a functor that $f_* g_* = \mathrm{Id}_{F(Y)}$ and $g_* f_* = \mathrm{Id}_{F(X)}$, which ensures that objects $F(X)$ and $F(Y)$ are isomorphic as well. □

Theorem 1 is of fundamental importance. Here is a standard way of applying it: suppose that we want to find out whether some given topological spaces $X$ and $Y$ are distinct. Take a functor from the category of topological spaces to another category, for instance, a category of groups, and compare the objects $F(X)$ and $F(Y)$. If they are distinct, then $X$ and $Y$ are distinct as well. In case $F(X)$ and $F(Y)$ coincide, nothing can be said about $X$ and $Y$. This remark explains the significance of homology theory, which is a functor from topology to algebra.

**Exercise 2.** Applying Theorem 1, show that the cyclic groups $Z_4$ and $Z_5$ are not isomorphic.

Thus, with the help of a functor $F\colon \mathsf{G}_1 \to \mathsf{G}_2$, the problem of distinguishing objects in the category $\mathsf{G}_1$ is replaced by a similar problem of distinguishing objects in the category $\mathsf{G}_2$. The meaning of the replacement is that in $\mathsf{G}_2$ this problem may be easier. It should be noted that when we pass from the category $\mathsf{G}_1$ to $\mathsf{G}_2$, a part of the information about the objects of $\mathsf{G}_1$ is usually lost.

A careful consideration of the above arguments shows that a "nice" functor should possess the following properties:

1. It should be easily computable, *i.e.* the determination of the object $F(X)$ for a given space $X$ should not pose difficulties of fundamental nature.
2. There should be a simple way of distinguishing objects $F(X)$ and $F(Y)$.
3. The transition from an object $X$ to $F(X)$ should not lose too much information.

Homology functors from the category of topological spaces to the category of groups meet these requirements to a significant extent. These aspects should be given special attention when studying homology theory.

In fact, a homology functor assigns to a topological space not a single group but rather a whole sequence of Abelian groups, *i.e.* it is a functor to the category of sequences of Abelian groups. Calculating the homology groups of an arbitrary space may prove to be unpredictably difficult, therefore, as a rule, they are studied for a class of spaces that are not too complicated. We take the category of simplicial complexes as the domain of our homology functor and only briefly mention more general homology theories.

## 1.2 Some geometric properties of $\mathbb{R}^N$

Recall that a basis in the Euclidean space $\mathbb{R}^N$ is an ordered collection of $N$ linearly independent vectors.

**Definition.** Two bases of $\mathbb{R}^N$ are called *equivalent* if the determinant of any change of coordinates matrix between them is positive.

**Exercise 3.** Prove that the relation thus introduced is an equivalence relation.

Being an equivalence relation, the above relation on the set of all bases decomposes it into two classes of equivalent bases.

**Example.** How do we decide, without calculating the matrix, whether two bases of the line (of the plane, or of the 3-space) are equivalent? We assume that the bases are given by a picture. The answer is simple. Two bases of the line (*i.e.* two vectors) are equivalent if they are co-directed. To each basis of the plane we can assign the rotation from the first vector to the second one along the smaller angle. Two bases are equivalent if either both of them are positive (*i.e.* are counterclockwise) or both are negative (clockwise). Finally, all bases of the 3-space can be decomposed into the left ones and the right ones, depending on whether the rotation from the first vector to the second one in the direction of the smaller angle is positive or negative when looked upon from the end of the third vector. Two bases are equivalent if they are of the same type. To determine the type of a basis, one could use the physicists' "screwdriver rule".

**Exercise 4.** Figure 1 shows three right bases and a left one. Find the left one.

**Definition.** *Orientation* of the space $\mathbb{R}^N$ is a class of equivalent bases.

Orientation is usually given by specifying a basis representing the relevant equivalence class.

**Exercise 5.** Prove that $\mathbb{R}^N$ has precisely two distinct orientations.

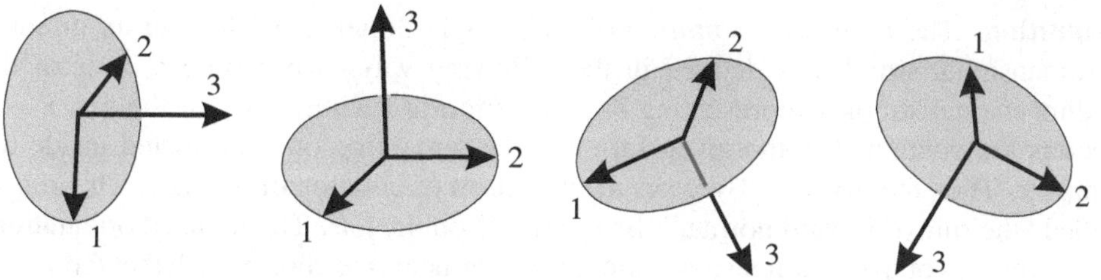


Figure 1. Right bases and a left one in $\mathbb{R}^N$.

It is also convenient to stipulate that the space $\mathbb{R}^0$ (the point) has two orientations, the orientation "+" and the orientation "–".

**Definition.** A system $a_0, a_1, \ldots, a_n$ of $n+1$ points in $\mathbb{R}^N$ is called *independent* if these points are not contained in the same plane of dimension $n-1$ (or less).

We would like to stress that any system of $n$ points is contained in some plane of dimension $\leqslant n-1$.

**Exercise 6.** Prove that the independence of points $a_0, a_1, \ldots, a_n$ is equivalent to the linear independence of the vectors $\overline{a_0a_1}, \overline{a_0a_2}, \ldots, \overline{a_0a_n}$.

**Exercise 7.** Prove that any subset of an independent system of points is also an independent system of points.

**Definition.** The convex hull of $n+1$ independent points $a_0, a_1, \ldots, a_n$ in $\mathbb{R}^N$ is called an *n-dimensional simplex*. The points $a_0, a_1, \ldots, a_n$ are called the *vertices* of the simplex.

It follows from the definition that simplices of dimension 0, 1, 2, and 3 are points, segments, triangles, and tetrahedra, respectively.

The plane of the smallest dimension containing a given simplex is called the *support plane* of that simplex. Its dimension coincides with that of the simplex. *Orientation* of a simplex is an orientation of its support plane. It is given by the choice of a basis. According to our agreement on the orientations of $\mathbb{R}^N$, a 0-dimensional simplex, *i.e.* a point, has two possible orientations, "+" and "–".

**Definition.** A *face* of a simplex is the convex hull of some subset of the set of its vertices.

**Exercise 8.** Prove that a face of a simplex is itself a simplex.

**Exercise 9.** How many $m$-dimensional faces does an $n$-dimensional simplex have?

**Exercise 10.** What is the total number of faces of an $n$-dimensional simplex?

**Definition.** The *induced orientation* of an $(n-1)$-dimensional face of an oriented $n$-dimensional simplex is defined in the following way: we choose a basis of the $n$-dimensional simplex representing its orientation in such a way that the first $n-1$ vectors are contained in the given face and the remaining one is directed inside the simplex. Then the first $n-1$ vectors determine an orientation of the face. This rule is called "the rule of inward normal". See Figure 2 on the left. The induced orientations of the vertices of a one-dimensional simplex (a segment) are chosen such that the vector that orients the segment is directed from the plus to the minus.

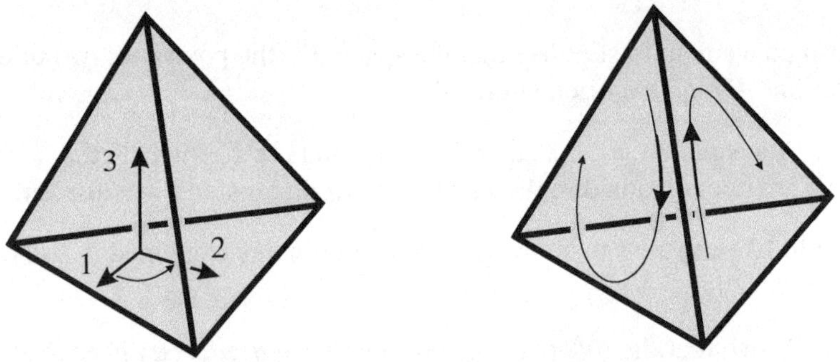


Figure 2. The induced orientation is defined by the rule of "inward normal". The iteration of this rule yields opposite orientations.

If $\alpha$ is the orientation of a simplex $\sigma$ and $\delta$ is a face of it, then the induced orientation of $\delta$ is denoted as $\alpha|\delta$.

**Theorem 2** (On doubly-induced orientations). *Let an $(n-2)$-dimensional simplex $\gamma$ be a common face of $(n-1)$-dimensional faces $\delta_1$, $\delta_2$ of an $n$-dimensional simplex $\sigma$ with orientation $\alpha$. Then the orientations $(\alpha|\delta_1)|\gamma$ and $(\alpha|\delta_2)|\gamma$ are opposite.*

The proof of this theorem is obtained by a direct application of the definition of the induced orientation. Therefore we omit it, restricting ourselves to the illustration in Figure 2, right.

**Definition.** A finite collection of simplices in $\mathbb{R}^N$ is called a *simplicial complex* if any two of its simplices either have no common points or intersect along their common face.

We can stipulate that simplices without common points intersect along their common empty face. Then the above definition can be reduced to requiring that any two simplices intersect along their common face. We emphasize that, from the formal point of view, it is necessary to distinguish the notion of a simplicial complex (a collection of simplices) and of its *underlying space* (the union of these simplices). The underlying topological space of $K$ is denoted by $|K|$. It is always a *polyhedron*, *i.e.* it can be presented as the union of some convex polytopes in $\mathbb{R}^N$. In this situation we say that the complex $K$ *triangulates* the polyhedron $|K|$ (or represents a *triangulation* of it). The *dimension* of $K$ is defined as the maximal dimension of its simplices.

**Exercise 11.** Give examples of simplicial complexes in the plane and in 3-space, as well as an example of a collection of simplices that does not form a simplicial complex.

**Definition.** An *orientation* of a simplicial complex is a set of orientations of each of its simplices including their faces.

**Exercise 12.** How many distinct orientations does the triangle, viewed as a simplicial complex, have?

The construction of the homology groups of a simplicial complex is carried out in two steps: first, to each simplicial complex we assign a certain so-called chain complex, then to this chain complex we assign its homology groups. From the methodological point of view it is more convenient to start with the second step.

## 1.3 Chain complexes

**Definition.** A sequence $C$ of Abelian groups and their homomorphisms

$$\cdots \longrightarrow C_{n+1} \xrightarrow{\partial_{n+1}} C_n \xrightarrow{\partial_n} C_{n-1} \longrightarrow \cdots,$$

infinite in both directions, is called a *chain complex* if for all $n$ we have the equality $\partial_n \partial_{n+1} = 0$.

Let us stress that the equality $\partial_n \partial_{n+1} = 0$ should be understood in the following way: for any element $x$ of the group $C_{n+1}$ the element $\partial_n(\partial_{n+1}(x))$ should be the trivial element of $C_{n-1}$. We denote it by zero, since we employ the additive notation for the chain groups.

**Definition.** The group $C_n$ is called the *$n$-dimensional chain group* of the complex $C$. The kernel $\operatorname{Ker} \partial_n \subset C_n$ of the homomorphism $\partial_n$ is called the *group of $n$-dimensional cycles* and is denoted by $A_n$. The image $\operatorname{Im} \partial_{n+1} \subset C_n$ of $\partial_{n+1}$ is called the *group of $n$-dimensional boundaries* of $C$ and is denoted by $B_n$.

**Exercise 13.** Give an example of a sequence of groups and their homomorphisms that is not a chain complex.

**Exercise 14.** Find the groups of cycles and the groups of boundaries for all dimensions of the complex

$$\cdots \longrightarrow 0 \longrightarrow \mathbb{Z} \oplus \mathbb{Z} \xrightarrow{\partial_2} \mathbb{Z} \longrightarrow 0 \longrightarrow \cdots,$$

where the chain groups are given by the equalities $C_1 = \mathbb{Z}$, $C_2 = \mathbb{Z} \oplus \mathbb{Z}$, $C_n = 0$ for $n \neq 1, 2$, and the homomorphism $\partial_2$ is defined by the rule $\partial_2(m, n) = 3m + 3n$.

It is easy to show that for any chain complex the group of boundaries $B_n$ is contained in the group of cycles $A_n$. The inverse is also true; if $B_n \subset A_n$ for all $n$, then the given sequence of groups and their homomorphisms is a chain complex, *i.e.* $\partial_n \partial_{n+1}$ is always 0.

**Definition.** The quotient group $A_n/B_n$ is called the *n-dimensional homology group* of the chain complex $C$ and is denoted by $H_n(C)$.

**Exercise 15.** Calculate the homology groups of the complex of Exercise 14.

**Terminology.** The elements of the group $A_n$ are called *cycles* and those of $B_n$ are called *boundaries*. The homomorphisms $\partial_n$ are called *boundary homomorphisms*. Two cycles $a_1, a_2 \in A_n$ are called *homologous* if their difference $a_1 - a_2$ is a boundary, *i.e.* is an element of $B_n$. Thus, two cycles determine the same element of the homology group if and only if they are homologous. The elements of each homology group can be interpreted as classes of homology equivalent cycles.

**Exercise 16.** Calculate the homology groups of the elementary complex $E(m)$ which has the form

$$\cdots \longrightarrow 0 \xrightarrow{\partial_{m+1}} \mathbb{Z} \xrightarrow{\partial_m} 0 \longrightarrow \cdots$$

and whose chain groups are the following:

$$E_n(m) = \begin{cases} 0, & n \neq m, \\ \mathbb{Z}, & n = m. \end{cases}$$

**Exercise 17.** Calculate the homology groups of the elementary complex $D(m, k)$ which has the form

$$\cdots \longrightarrow 0 \longrightarrow \mathbb{Z} \xrightarrow{\partial_{m+1}} \mathbb{Z} \longrightarrow 0 \longrightarrow \cdots$$

and whose chain groups are given by the rule

$$D_n(m, k) = \begin{cases} 0, & n \neq m, m+1, \\ \mathbb{Z}, & n = m, m+1, \end{cases}$$

and the homomorphism $\partial_{m+1}$ consists in multiplication by an integer $k \neq 0$.

**Exercise 18.** Give a definition of the direct sum of chain complexes and prove that $H_n(C \oplus C') = H_n(C) \oplus H_n(C')$.

**Definition.** Let $C$ and $C'$ be two chain complexes. A family of homomorphisms $\varphi = \{\varphi_n : C_n \to C'_n, -\infty < n < \infty\}$ is called a *chain map* if $\varphi_n \partial_{n+1} = \partial_{n+1} \varphi_{n+1}$ for all $n$.

The meaning of the condition $\varphi_n \partial_{n+1} = \partial_{n+1} \varphi_{n+1}$ is that all the squares in the diagram

$$\begin{array}{ccccccccc} \cdots & \longrightarrow & C_{n+1} & \xrightarrow{\partial_{n+1}} & C_n & \xrightarrow{\partial_n} & C_{n-1} & \longrightarrow & \cdots \\ & & \downarrow{\scriptstyle \varphi_{n+1}} & & \downarrow{\scriptstyle \varphi_n} & & \downarrow{\scriptstyle \varphi_{n-1}} & & \\ \cdots & \longrightarrow & C'_{n+1} & \xrightarrow{\partial_{n+1}} & C'_n & \xrightarrow{\partial_n} & C'_{n-1} & \longrightarrow & \cdots \end{array}$$

are commutative.

**Exercise 19.** Let $\varphi\colon C \to C'$ be a chain map. Prove that $\varphi_n(A_n) \subset A'_n$ and $\varphi_n(B_n) \subset B'_n$, *i.e.* that $\varphi$ takes cycles to cycles and boundaries to boundaries.

**Theorem 3.** *Let $\varphi\colon C \to C'$ be a chain map between chain complexes. Then for any integer $n$ assigning to each cycle $x \in C_n$ the chain $\varphi_n(x) \in C'_n$ induces a well-defined homomorphism $\varphi_*\colon H_n(C) \to H_n(C')$.*

*Proof.* This theorem is almost obvious. Its proof does not present any difficulties, especially if the reader has completed Exercise 19. Nevertheless let us describe explicitly how a chain map $\varphi$ between two chain complexes induces homomorphisms $\varphi_*$ between the homology groups of matching dimensions. Here we encounter for the first time the so-called *diagrammatic search* (which is our preferred way to refer to the approach that is also known as "general nonsense"). This method is just a collection of some more or less standard tricks applied to diagrams. It is best to observe the method in practice. Let us reproduce the above diagram having removed, for simplicity, all the indices, see Figure 3. The elements used in the process of the proof are placed next to the groups to which they belong.

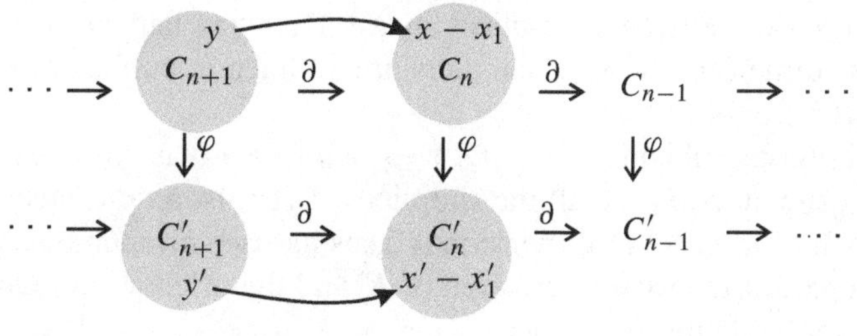


Figure 3. A proof via the method of diagrammatic search.

Let $h$ be an arbitrary element of $H_n(C)$. We would like to use the given chain map $\varphi$ to assign to $h$ a well-defined element $h' = \varphi_*(h)$ of $H_n(C')$. The first step consists in choosing some cycle $x \in A_n \subset C_n$ representing $h$. Applying to $x$ the homomorphism $\varphi$, we obtain a certain element $x' = \varphi(x)$ of the group $C'_n$.

Let us show that $x'$ is a cycle. Indeed, $\partial x' = \partial\varphi(x) = \varphi(\partial x) = \varphi(0) = 0$ (we have used the commutativity of the diagrams and the fact that $x$ itself is a cycle). Now we can define $h'$ as the class containing the cycle $x'$.

Let us prove that the element $h'$ of the group $H_n(C')$ does not depend on $x$, *i.e.* on the choice of an element representing $h$. Let $x_1$ be another representative, and let $x_1' = \varphi(x_1)$ be its image in $C_n'$. Denote by $h''$ the equivalence class containing $x_1'$. Then the difference $x - x_1$ is a boundary. Therefore there exists an element $y$ of $C_{n+1}$ such that $\partial y = x - x_1$. Again using the commutativity, we obtain that $x' - x_1' = \varphi(x - x_1) = \varphi\partial(y) = \partial\varphi(y) = \partial y'$, where $y' = \varphi(y)$. It follows that $x'$ and $x_1'$ differ by a boundary element, which means that the elements $h'$ and $h''$ coincide. □

**Exercise 20.** Describe the category of all chain complexes and the category of sequences of Abelian groups. Check that assigning to each chain complex the sequence of its homology groups, and assigning to each chain map $\varphi$ between chain complexes the induced map $\varphi_*$ between their homology groups, yield together a functor from the former of the above two categories to the latter.

## 1.4 Homology groups of a simplicial complex

Let $K$ be an oriented simplicial complex. We assign to it a chain complex $C(K)$ as follows. The elements of the $n$-dimensional chain group $C_n(K)$ are formal linear combinations of the form $m_1\sigma_1 + m_2\sigma_2 + \cdots + m_k\sigma_k$, where $m_i$ are integers and $\sigma_1, \ldots, \sigma_k$ are all the $n$-dimensional simplices. The addition is coordinate-wise. Of course, the set $C_n(K)$ is a group with respect to this operation.

From the algebraic point of view, $C_n(K)$ is the free Abelian group that is freely generated, in an obvious sense, by the set of all the $n$-dimensional simplices. In particular, its rank is equal to the number of these simplices. Furthermore, suppose that there are no $n$-dimensional simplices in $K$. This may happen if $n$ is negative or greater than the dimension of $K$. Then there are no linear combinations either. In this case we set $C_n(K) = 0$.

To define homomorphisms $\partial_n \colon C_n(K) \to C_{n-1}(K)$ it is sufficient to define the images of the generators, *i.e.* of all the simplices. Let $\sigma$ be an $n$-dimensional simplex of $K$. Then each of its $(n-1)$-dimensional faces has two orientations, its own orientation that is a part of the total orientation of $K$ and the orientation induced on it as a face of $\sigma$. We set by definition

$$\partial_n(\sigma) = \sum_{\delta_i \in K} \varepsilon_i \delta_i,$$

where the summation is over all the simplices $\delta_i$ of dimension $n-1$ and the numbers $\varepsilon_i$ (called the *incidence coefficients*) are given by the following rule:

$$\varepsilon_i = \begin{cases} 0, & \text{if } \delta_i \text{ is not a face of } \sigma; \\ 1, & \text{if } \delta_i \text{ is a face of } \sigma \text{ and the two orientations coincide;} \\ -1, & \text{if } \delta_i \text{ is a face of } \sigma \text{ and the two orientations are distinct.} \end{cases}$$

The geometric meaning of this rule is quite simple. Recall that $\sigma$ denotes not just an oriented simplex but also the chain $1 \cdot \sigma$ (an element of $C_n(K)$). Let us postulate that the chain $-\sigma = (-1) \cdot \sigma$ corresponds to the same simplex $\sigma$, but taken with the opposite orientation. Then $\partial_n(\sigma)$ is nothing more than the boundary of $\sigma$, where all the $(n-1)$-dimensional simplices contained in it are taken with their induced orientations.

**Theorem 4.** *For any simplicial complex $K$ the groups $C_n(K)$ and the homomorphisms $\partial_n : C_n(K) \to C_{n-1}(K)$ form a chain complex (which we denote by $C(K)$).*

The proof of this theorem follows from Theorem 2 on doubly-induced orientations, which we now can rephrase as follows: the boundary of a boundary is empty. The fundamental importance of this fact, which lies in the foundation of any homology theory, merits a careful and deep consideration.

**Definition.** Let $K$ be an oriented simplicial complex. Then the homology groups of the corresponding chain complex $C(K)$ are called the *homology groups of $K$* and are denoted by $H_n(K)$.

In other words, the group $H_n(K)$ is the quotient group $\operatorname{Ker} \partial_n / \operatorname{Im} \partial_{n+1}$ of the kernel of the homomorphism $\partial_n$ by the image of the homomorphism $\partial_{n+1}$.

**Exercise 21.** Prove that the groups $H_n(K)$ do not depend on the choice of an orientation of $K$.

One can also prove that the homology groups of any *polyhedron* (a subset of $\mathbb{R}^N$ which can be presented as a simplicial complex) do not depend on any particular choice of such presentation, *i.e.* on the triangulation. The proof of this result is rather cumbersome, although it does not present serious difficulties of theoretical nature. For instance, one may proceed as follows:

1. Make sure that the above construction of the homology groups can be carried over to polyhedra decomposed not necessarily into simplices but into arbitrary polytopes.

2. Prove that if some decomposition of a triangulated polyhedron $K$ into polytopes has the property that each simplex consists of whole polytopes, then the homology groups calculated via the triangulation are isomorphic to those calculated via the decomposition into polytopes. It is easiest to describe the desired isomorphism using the generators of the chain groups of the triangulation, *i.e.* simplices. To each $n$-dimensional simplex $\sigma$ of the triangulation we assign the chain that consists of all the polytopes comprising $\sigma$. The coefficients at those polytopes are equal to $\pm 1$, depending on whether the orientation of a given polytope coincides with that of $\sigma$ or is opposite to it. Therefore the boundary of this chain coincides with the boundary of the simplex, which essentially ensures the isomorphism of the homology groups, see also Figure 4.

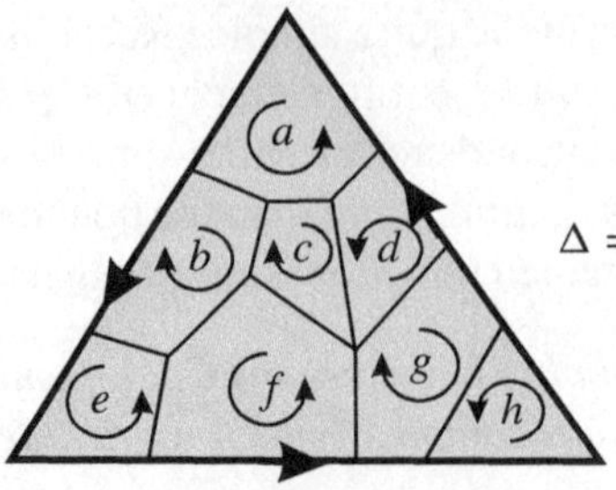


Figure 4. An oriented simplex is assigned the chain composed of the polytopes comprising it.

3. For any two given triangulations of the same polyhedron we may consider the decomposition into the polytopes given by intersections of any two simplices belonging to the two triangulations. Then the previous item ensures the desired isomorphism of the homology groups.

It is important to mention that it is possible to define the homology groups not only for polyhedra but also for more general spaces homeomorphic to polyhedra. Sometimes such spaces are called *topological polyhedra.* Topological polyhedra can by definition be triangulated, but into *curvilinear simplices* (images of genuine simplices under the relevant homeomorphism).

Thus, in order to calculate the homology groups of a given topological space, one should perform the following steps:

1. Present the space as a polyhedron and triangulate it.

2. Choose an orientation for the simplicial complex thus obtained.

3. Calculate the chain groups $C_n$.

4. Describe the boundary homomorphisms $\partial_n$.

5. Calculate the groups of cycles $A_n$.

6. Calculate the groups of boundaries $B_n$.

7. Calculate the quotient groups $H_n = A_n/B_n$.

**Theorem 5.** *The homology groups of the point are the following:*

$$H_n(*) = \begin{cases} 0, & n \neq 0, \\ \mathbb{Z}, & n = 0. \end{cases}$$

*Proof.* The proof is evident, since the chain complex corresponding to the point (viewed as a 0-dimensional simplex) has the form,

$$\cdots \longrightarrow 0 \longrightarrow \mathbb{Z} \longrightarrow 0 \longrightarrow \cdots,$$

where $\mathbb{Z}$ is the 0th chain group. □

**Exercise 22.** Calculate the homology groups of the segment and those of the circle.

**Exercise 23.** Calculate the groups $H_2(S^2)$, $H_2(T^2)$, where $S^2$ is the two-dimensional sphere and $T^2 = S^1 \times S^1$ is the two-dimensional torus.

**Exercise 24.** Prove that for any simplicial complex $K$ the group $H_0(K)$ is the free Abelian group of rank equal to the number of connected components of $K$.

## 1.5 Simplicial maps

**Definition.** A map from one simplex to another is called *linear* if it takes vertices to vertices and can be extended to an affine map between the support planes of the simplices.

A linear map can of course take a simplex to a simplex of smaller dimension. For instance, a simplex can be projected onto one of its faces, see Figure 5.

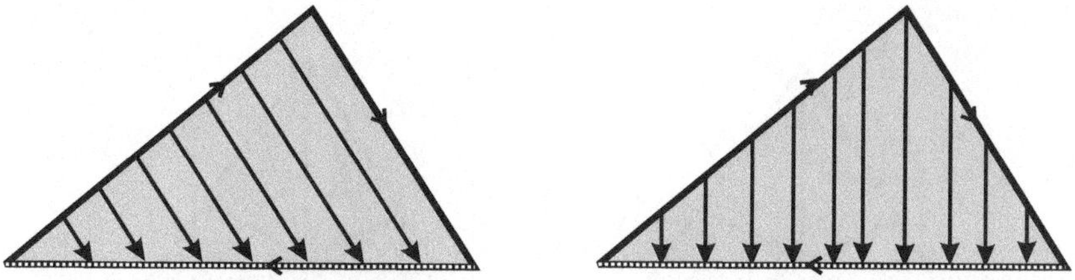

Figure 5. Linear (left) and non-linear (right) maps from a simplex onto its face.

**Exercise 25.** Prove that a linear map between two simplices is completely determined by the images of the vertices.

**Definition.** A map $f\colon K \to L$ between simplicial complexes (more precisely, a map $f\colon |K| \to |L|$) is called *simplicial* if the image of each simplex of $K$ is a simplex of $L$ and the restriction of $f$ to any simplex of $K$ is a linear map.

**Exercise 26.** Suppose that a map $f^{(0)}\colon K^{(0)} \to L^{(0)}$ takes the vertex set $K^{(0)}$ of $K$ to the vertex set $L^{(0)}$ of $L$. Prove that $f^{(0)}$ can be extended to a simplicial map $f\colon K \to L$

if and only if it has the following property: if vertices $v_0, v_1, \ldots, v_m$ belong to some simplex of $K$, then also the vertices $f^{(0)}(v_0), f^{(0)}(v_1), \ldots, f^{(0)}(v_m)$ belong to some simplex of the complex $L$.

Each simplicial map $f: K \to L$ between oriented simplicial complexes gives rise to a map $\varphi: C(K) \to C(L)$ between the corresponding chain complexes, which is defined by the images of the generators (*i.e.* of simplices) via the formula:

$$\varphi(\sigma) = \begin{cases} 0, & \text{if } \dim \sigma > \dim f(\sigma); \\ f(\sigma), & \text{if } \dim \sigma = \dim f(\sigma) \text{ and } f_{|\sigma} \text{ is orientation-preserving}; \\ -f(\sigma), & \text{if } \dim \sigma = \dim f(\sigma) \text{ and } f_{|\sigma} \text{ is orientation-reversing}. \end{cases}$$

By $f_{|\sigma}$ we denote the restriction of $f$ to the simplex $\sigma$. Since this restriction is an affine isomorphism in our case, it makes sense to say whether $f$ preserves or reverses the orientation.

**Exercise 27.** Prove that $\varphi$ is a chain map.

**Definition.** A complex $K_1$ is called a *subdivision* of a complex $K$ if their underlying spaces coincide as subsets of $\mathbb{R}^N$ and each simplex of the complex $K_1$ is contained in some simplex of the complex $K$.

Subdivisions are usually obtained by decomposing each simplex into some smaller ones in such a way that the decompositions are consistent on all the faces.

**Example.** Figure 6 shows the *barycentric* and the *second barycentric* subdivisions of a triangle.

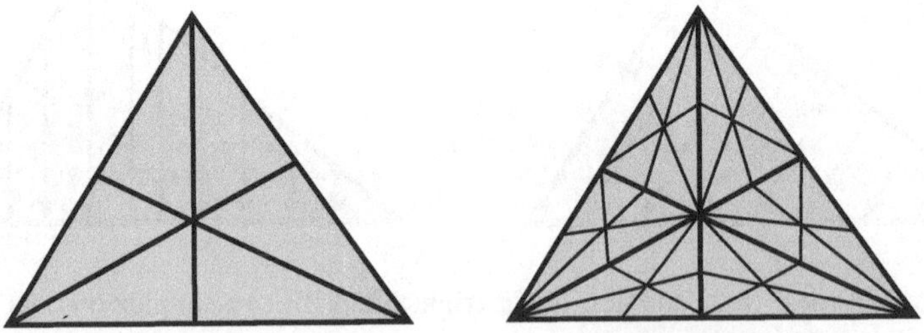

Figure 6. The barycentric and the second barycentric subdivisions of a triangle.

**Exercise 28.** Prove that each complex has an arbitrarily small subdivision.

**Theorem 6** (on simplicial approximation). *For any continuous map $f: |K| \to |L|$ between the underlying spaces of two simplicial complexes there exists a subdivision $K_1$ of the complex $K$ and a simplicial map $g: K_1 \to L$ such that $g$ is an approximation of $f$ in the following sense: for any point $x \in |K|$ there is a simplex of the complex $L$ that contains both $f(x)$ and $g(x)$.*

The use of the term "approximation" is due to the following fact. If a subdivision of $L$ is chosen to be so small that the diameters of all simplices are bounded from above by some chosen $\varepsilon$, then the map $g$ is an $\varepsilon$-approximation of $f$ in the metric sense. On the other hand, forcing each point $f(x)$ to move steadily towards the point $g(x)$ along some segment that joins these points and is contained inside the simplex, we get a continuous deformation of $f$ into $g$. Thus, $g$ is an approximation of $f$ in the homotopy sense. Let us state this precisely.

**Definition.** Maps $f, g\colon X \to Y$ from one topological space to another one are called *homotopic* (this is denoted as $f \sim g$) if there exists a continuous map $F\colon X \times I \to Y$ such that $F(x, 0) = f(x)$ and $F(x, 1) = g(x)$ for all $x$ (here and henceforth $I$ denotes the segment [0,1]).

If $g$ is a simplicial approximation of a map $f\colon |K| \to |L|$, then a homotopy $F\colon |K| \times I \to |L|$ between them can be given by the formula $F(x,t) = tg(x) + (1-t)f(x)$, which is well defined because the points $f(x)$, $g(x)$ belong to the same simplex.

The proof of the simplicial approximation theorem makes use of the following definitions and facts.

**Definition.** The *open star* $\overset{\circ}{\mathrm{St}}(v, K)$ of a vertex $v$ of a complex $K$ is the union of interiors of all the simplices of $K$ of which $v$ is a vertex. If we take the union of the closed simplices, then we get the *closed star* $\mathrm{St}(v, K)$.

Examples of open stars are shown in Figure 7.

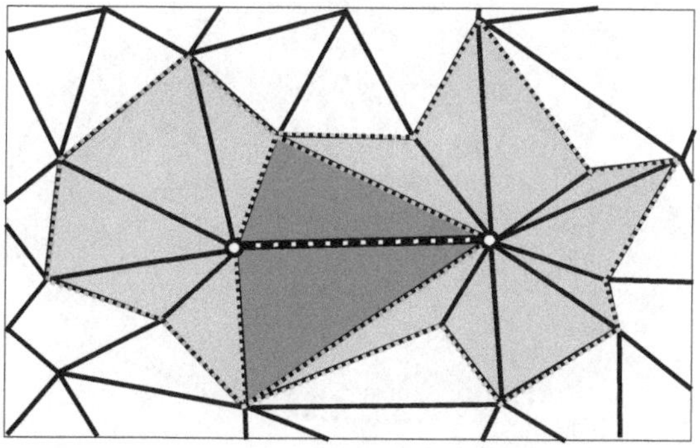

Figure 7. The open stars of two vertices joined by an edge have non-empty intersection.

**Exercise 29.** Prove that the open star of any vertex $v$ of a complex $K$ is an open subset of the underlying space of $K$.

**Exercise 30.** Let $v_1, \ldots, v_k$ be some vertices of a complex $K$. Prove that the intersection $\bigcap_{i=1}^{k} \mathring{\mathrm{St}}(v_i, K)$ of their open stars is non-empty if and only if $K$ contains a simplex spanning these vertices (see the illustration in Figure 7).

**Definition.** A number $\delta > 0$ is called a *Lebesgue number* of a cover $\{U_\alpha\}$ of a metric space $X$ ($\alpha$ runs over some indexing set $A$) if for any $Y \subset X$ such that the diameter of $Y$ does not exceed $\delta$ there exists $\alpha \in A$ such that $Y \subset U_\alpha$.

**Exercise 31.** Prove that for any open cover of a metric compact space there exists a Lebesgue number.

Let us now prove the simplicial approximation theorem.

*Proof.* Let $f\colon |K| \to |L|$ be an arbitrary continuous map between the underlying spaces of the simplicial complexes $K$ and $L$. For each vertex $w$ of the complex $L$ we define an open set $U_w = f^{-1}(\mathring{\mathrm{St}}(w, L)) \subset K$. Then the family of subsets $\{U_w\}$, where $w$ runs over the set of all vertices of $L$, is an open cover of the polyhedron $|K|$. Let $\delta$ be a Lebesgue number of this cover, and let $K_1$ be a sufficiently small subdivision of $K$ such that the diameter of the star of any vertex of $K_1$ does not exceed $\delta$. Let us assign to each vertex $v$ of the complex $K_1$ a vertex $w$ of the complex $L$ such that $\mathring{\mathrm{St}}(v, K_1) \subset U_w$, see Figure 8. The vertex $w$ exists, since the diameter of the star $\mathring{\mathrm{St}}(v, K_1)$ is not greater than the Lebesgue number. If there are several such vertices, then we just take any one of them. This assignment defines a map $g^0\colon K_1^0 \to L^0$ of the vertex set of the complex $K_1$ to that of the complex $L$.

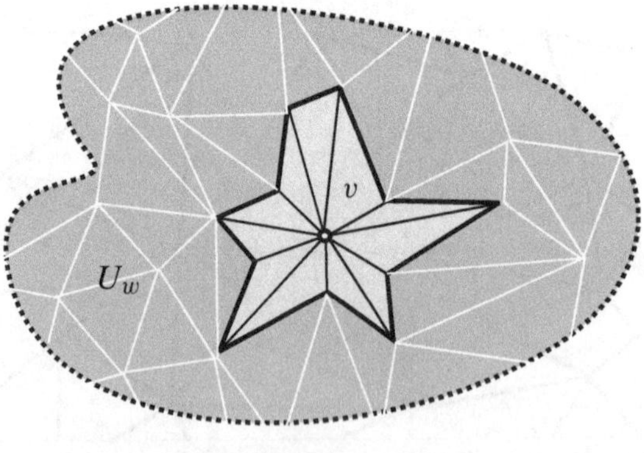


Figure 8. The star of each vertex $v \in K_1$ is contained in the pre-image of the open star of some vertex $w \in L$.

Let us prove that the map $g^0$ has the following property: if points $v_1, \ldots v_k$ span a simplex in $K_1$, then their images $w_1 = g^0(v_1), \ldots, w_k = g^0(v_k)$ (some of which may coincide) also span a simplex in $L$. Indeed, since $v_1, \ldots v_k$ span a simplex, their stars have non-empty intersection, hence the stars of $w_1, \dots w_k$ in $L$ also have non-empty intersection. Therefore they also span a simplex. This guarantees the existence of the simplicial map $g\colon K_1 \to L$ that extends $g_0$. Obviously, $g$ approximates $f$ in the sense that for any point $x \in |K|$ its images $f(x)$ and $g(x)$ lie in the same simplex. □

The relative version of the simplicial approximation theorem is also true.

**Theorem 7** (on the relative simplicial approximation). *Consider simplicial complexes $K, L$, their subcomplexes $M \subset K$, $N \subset L$ and a map $f\colon |K| \to |L|$ such that $f(|M|) \subset |N|$. Then there exists a subdivision $K_1$ of the complex $K$ and a simplicial map $g\colon K_1 \to L$ such that for any point $x \in |K|$ its images $f(x)$ and $g(x)$ lie in the same simplex of $L$ and for any point $x \in |M|$ its images $f(x)$ and $g(x)$ lie in the same simplex of the subcomplex $N$.*

**Exercise 32.** Prove the relative version of the simplicial approximation theorem.

## 1.6 Induced homomorphisms of homology groups

Being in possession of the simplicial approximation theorem and knowing how a simplicial map induces homomorphisms of homology groups, it is not difficult to define the induced homomorphisms $f_*\colon H_n(K) \to H_n(L)$ for an arbitrary map $f\colon |K| \to |L|$.

**Definition.** Let $f\colon P \to Q$ be an arbitrary map between polyhedra, and let $g\colon K \to L$ be any simplicial map of some complexes defining triangulations of $P$ and $Q$, homotopic to $f$ (such a $g$ exists by the simplicial approximation theorem). Then we define homomorphisms $f_*\colon H_n(P) \to H_n(Q)$ by setting $f_* = g_*$ for all $n$.

It is necessary to ensure that the maps $f_*$ are well defined, *i.e.* to prove that if $g'\colon K' \to L$ is another simplicial approximation of $f$, then $g_* = g'_*$. The idea of the proof is this: $f \sim g$ and $f \sim g'$ imply that $g \sim g'$, *i.e.* that there exists a map $F\colon P \times I \to Q$ such that $F(x, 0) = g(x)$ and $F(x, 1) = g'(x)$. By the relative simplicial approximation theorem applied to the map $F$ (which is simplicial on the upper and the lower bases of the cylinder $P \times \{0, 1\}$), this map can be replaced by a map $G\colon P \times I \to Q$ which is simplicial with respect to some triangulation of the polyhedron $P \times I$ and which takes $P \times \{0, 1\}$ to $g(K) \cup g'(K')$.

Now let $a$ be an arbitrary cycle in $K$. We want to prove that the cycles $g_*(a)$, $g'_*(a)$ are homologous, *i.e.* that their difference is the boundary of some chain. A geometric prototype of such a bounded chain is the image of the cylinder $a \times I$ under the homotopy $G$. Since $\partial(a \times I) = \partial a \times I \cup a \times \partial I = a \times \partial I = a \times \{0\} - a \times \{1\}$, we have that $\partial G(a \times I) = G\partial(a \times I) = G(a \times \{0\} - a \times \{1\}) = g_*(a) - g'_*(a)$. This actually means that the difference $g(a) - g'(a)$ is a boundary, *i.e.* the cycles $g_*(a)$ and $g'_*(a)$ determine the same element of the homology group, see Figure 9.

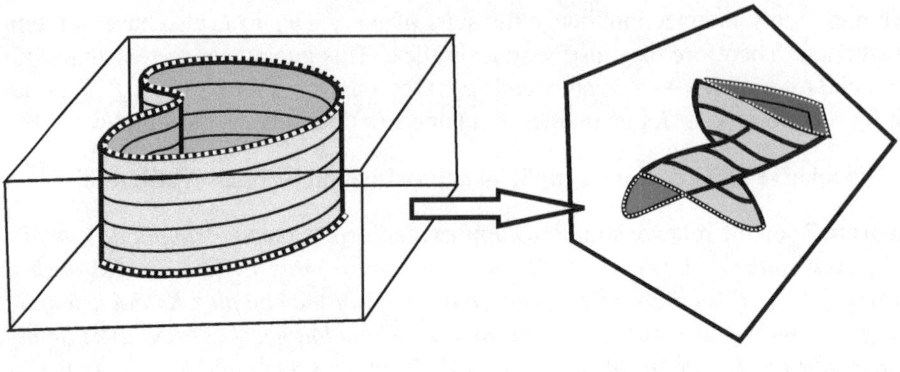

Figure 9. The "film" $G(a \times I)$ is bounded by the difference $g_*(a) - g'_*(a)$.

The idea just described allows us to better understand the formal proof (see, for instance, [6]), where it is explained, in particular, how one can triangulate the direct product $K \times I$, and why an algebraic object (the cycle $a$) can be treated as a geometric one, for instance, be multiplied by $I$.

**Exercise 33.** Prove that a composition of maps induces the composition of the homomorphisms and that the identity map induces the identity homomorphisms of the homology groups.

**Theorem 8.** *Let $f, g\colon P \to Q$ be homotopic maps between two polyhedra. Then the induced homomorphisms $f_*, g_*\colon H_n(P) \to H_n(Q)$ coincide for all $n$.*

The proof of the theorem follows directly from the definition of the induced homomorphisms. Indeed, any simplicial approximation of $f$ is at the same time a simplicial approximation of $g$, even if in the homotopy sense only, but it suffices.

**Exercise 34.** Prove that the homotopy equivalence of polyhedra $X$ and $Y$ implies that their homology groups are isomorphic.

## 1.7 Degrees of maps between manifolds

Recall that a topological space $M$ is called an *$n$-dimensional manifold* if every point $x \in M$ has a neighbourhood $U$ homeomorphic to a domain $V$ of the space $\mathbb{R}^n$. Any specific homeomorphism $\varphi\colon U \to V$ is called a *chart* on $M$. This chart can be viewed as a *local coordinate system*: the coordinates of a point $x \in U$ are exactly the coordinates of its image $\varphi(x)$ in $\mathbb{R}^n$. Any collection of charts whose domains cover the whole $M$ is called an *atlas*. If the domains $U$, $U'$ of two charts $\varphi\colon U \to V$ and $\psi\colon U' \to V'$ have a non-empty intersection $Z$, then there is a well-defined *change of charts* $\psi\varphi^{-1}\colon \varphi(Z) \to \psi(Z)$, see Figure 10.

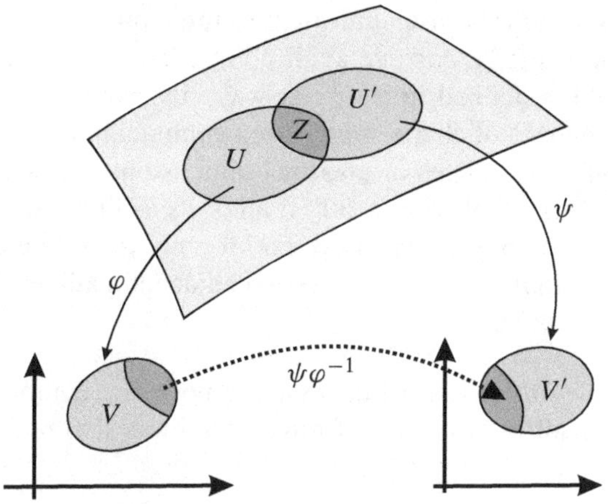


Figure 10. Changes of charts are homeomorphisms between domains in $\mathbb{R}^n$.

To avoid pathological examples it is normally required that the manifold be a Hausdorff topological space with a countable base. Manifolds are often considered endowed with extra structures. To determine such a structure, we should specify a class $\mathsf{C}$ of homeomorphisms between domains in $\mathbb{R}^n$ that is closed with respect to taking restrictions to smaller domains, taking inverse homeomorphisms, and taking a composition when it is well defined. Let us give three examples, where $\mathsf{C}$ consists of diffeomorphisms, piecewise-linear maps, and orientation-preserving maps.

1. A manifold is called *smooth* if it has an atlas in which all changes of charts are diffeomorphisms between domains in $\mathbb{R}^n$. Smooth manifolds have the advantage of being accessible to the powerful tools of mathematical analysis that exist for domains in $\mathbb{R}^n$. For instance, one can speak of smooth functions and smooth maps on smooth manifolds, singular and regular points, regular and critical values. All these notions are carried over from the case of $\mathbb{R}^n$ to the case of an arbitrary smooth manifold via introducing a local coordinate system in the neighbourhood of the point in question. The choice of a coordinate system is not important, since diffeomorphisms preserve the above properties.
2. A manifold is called *piecewise-linear* (or a *PL-manifold*) if it has an atlas in which all changes of charts are piecewise-linear (*i.e.* simplicial with respect to some triangulations) homeomorphisms between domains of $\mathbb{R}^n$. Since domains are non-compact, one has to consider infinite triangulations. It is known that a closed manifold is piecewise-linear if and only if it admits a triangulation such that the closed star of any vertex is piecewise-linear homeomorphic to the standard simplex.
3. A smooth manifold is called *orientable* if it admits an atlas such that all changes of charts preserve the orientation of the space $\mathbb{R}^n$. The property of preserving

the orientation is in this case equivalent to the requirement that the Jacobian of any change of charts be positive at all points. The orientability of a piecewise-linear manifold is defined in a similar way: the manifold must admit an atlas such that all changes of charts, which are simplicial homeomorphisms, preserve orientations of all simplices of maximal dimension. A triangulated manifold of dimension $n$ is orientable if and only if all of its $n$-dimensional simplices can be oriented in a coherent way, *i.e.* such that, for any $(n-1)$-dimensional simplex, the orientations induced on it by the two adjacent $n$-dimensional simplices are opposite to each other.

**Exercise 35.** Give definitions of a Euclidean, hyperbolic, complex-analytic, conformal, and Lipschitz manifold (to give a formal definition, it is in fact not necessary to understand the meaning of these terms).

**Theorem 9.** *For any connected closed triangulated manifold $M$ of dimension $n$ the group $H_n(M)$ is isomorphic to $\mathbb{Z}$ if $M$ is orientable, and is trivial if it is not.*

*Proof.* The sum $\sigma_1^n+\cdots+\sigma_m^n$ of all coherently oriented simplices of maximal dimension is a cycle. Since there are no $(n+1)$-dimensional simplices, there are no relations either, *i.e.* we obtain a non-trivial element of $H_n(M)$. All the other cycles of maximal dimension must be multiples of $\sigma_1^n+\cdots+\sigma_m^n$, because for any $n$-dimensional chain that defines a cycle, the coefficients at coherently oriented neighbouring simplices should be the same. For the same reason the absence of a coherent orientation ensures that the group $H_n(M)$ is trivial when $M$ is non-orientable. □

It is worth noting that if $M$ is disconnected, then $H_n(M) = \mathbb{Z}\oplus\cdots\oplus\mathbb{Z}$, where the number of summands $\mathbb{Z}$ coincides with the number of orientable connected components of $M$.

Thus, for any closed connected oriented manifold $M$ of dimension $n$ the group $H_n(M)$ is isomorphic to $\mathbb{Z}$ (with a canonic isomorphism determined by the orientation). Since each homomorphism $\mathbb{Z} \to \mathbb{Z}$ is essentially a multiplication by an integer, this allows us to introduce the notion of the degree of a map.

**Definition.** Let $f\colon M \to N$ be a simplicial map from one closed connected oriented triangulated manifold of dimension $n$ to another one, and let $f_*\colon H_n(M) \to H_n(N)$ be the induced homomorphism. Then the integer $f_*(1)$ is called the *degree* of the map $f$ and is denoted by $\deg f$.

The degree of a map admits a clear geometric interpretation. Let $\sigma$ be an arbitrary $n$-dimensional simplex of the target manifold $N$, and let $\sigma_1, \dots, \sigma_m$ be all the $n$-dimensional simplices from its pre-image, which $f$ maps to $\sigma$ via an affine isomorphism. Assume that the orientations of all the $\sigma_i$ and of $\sigma$ are consistent with those of the manifolds $M$ and $N$. Assign to each simplex $\sigma_i$ the number 1 if the restriction of $f$ to $\sigma_i$ preserves the orientation and $-1$ if $f$ reverses it, see Figure 11. Then the definition of the degree implies that it is equal to the sum of these numbers. Naturally, if $f$ is not simplicial, then prior to calculating its degree, $f$ should be approximated by a simplicial map.

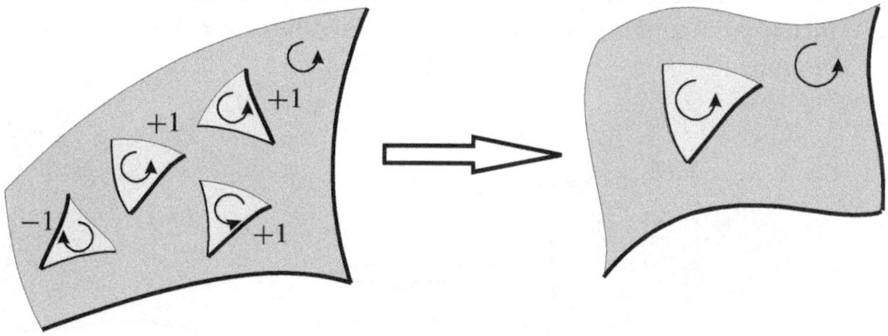


Figure 11. The degree of a map is equal to the difference between the number of the simplices mapped onto a given simplex $\sigma \subset N$ preserving the orientation and that of the simplices mapped onto $\sigma$ reversing the orientation.

**Exercise 36.** The circle $|z| = 1$ is mapped onto itself via the map $f$ given by the formula $f(z) = z^n$. Calculate the degree of this map.

The only difficulty which is likely to arise in the course of solving this exercise may be related to the necessity of passing to the simplicial category: both copies of the circle (the source and the target) must be replaced with simplicial complexes in such a way that $f$ is simplicial, see the solution of this exercise at the end of the book. Would it be possible to learn how to calculate the degrees of smooth maps between smooth manifolds without leaving the smooth category? It is, and sometimes this approach is more convenient.

Recall that a point $x \in M$ is called a *singular* point of a smooth map $f: M \to N$ from a smooth manifold $M$ of dimension $m$ to a smooth manifold $N$ of dimension $n \leqslant m$ if with respect to some (and hence to any) local coordinate systems $(x_1, \dots, x_m)$ in the neighbourhood of $x$ and $(y_1, \dots, y_n)$ in the neighbourhood of $y = f(x)$, the rank of the Jacobi matrix $(\partial y_i / \partial x_j)$ is less than $n$ (*i.e.* is not maximal). A point $y \in N$ is called a *regular value* if its pre-image does not contain singular points, and a *critical value* otherwise. According to the famous Sard theorem (see, for instance, [1]) the set of critical values of any smooth map between two manifolds is small. The rigorous meaning of the last word can be defined in different ways. For instance, one can say that this set has measure 0 or that it is of the first category (is the union of a countable family of nowhere dense sets). We will need just one property: that the complement of the set of critical values (*i.e.* the set of regular values) is always non-empty.

Let $f: M \to N$ be a smooth map from one closed oriented $n$-dimensional manifold to another one. Choose a regular value $b \in N$. Then the inverse function theorem and a compactness argument imply that the complete pre-image $f^{-1}(b) \subset M$ consists of a finite number of regular points $a_1, \dots, a_s \in M$. Assign to each point $a_k$ the number $\varepsilon_k = 1$ or $\varepsilon_k = -1$, according to the sign of the determinant of the Jacobi matrix $(\partial y_i / \partial x_j)$ (this determinant is non-zero, since the Jacobi matrix is a square one and has rank $n$), see Figure 12.

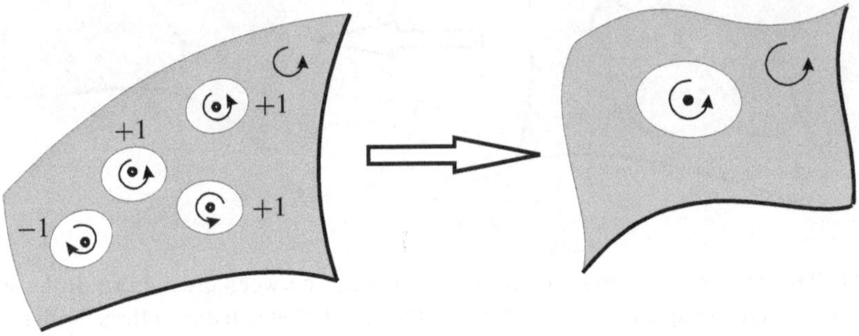


Figure 12. The degree of a smooth map is equal to the difference between the number of the points in the pre-image where the Jacobian is positive and the number of the points where the Jacobian is negative.

**Definition.** The number $\sum_{i=1}^{s} \varepsilon_i$, *i.e.* the difference between the number of the positive points and that of the negative points in the pre-image $f^{-1}(b) \subset M$, is called the *smooth degree* of the map $f$.

Evidently, the smooth degree coincides with the one introduced previously in terms of the induced homomorphism of the senior homology group, see the definition in Section 1.7. Intuitively this fact does not cause any doubts, it is sufficient to compare Figures 11 and 12. A rigorous proof can be obtained by passing to the category of piecewise-linear manifolds and their maps, where both methods of calculation are defined.

The equality between the degrees implies that the smooth degrees of homotopic smooth maps are equal. Let us give a useful geometric illustration of this fact. Let $f_0, f_1: M \to N$ be two homotopic maps from one closed oriented $n$-dimensional manifold to another one, and let $F: M \times I \to N$ be a smooth homotopy between these maps. Choose a regular value $b \in N$ of the map $F$. The implicit function theorem tells us that the pre-image $F^{-1}(b)$ consists of smooth arcs with the ends at the bases $N \times \{0, 1\}$ of the cylinder $N \times I$ and possibly of several smooth closed arcs inside of it. Furthermore, the ends of the same arc have the same sign if they belong to different bases of the cylinder and the opposite signs if they belong to the same base, see Figure 13. This ensures the equality $\deg(f_0) = \deg(f_1)$.

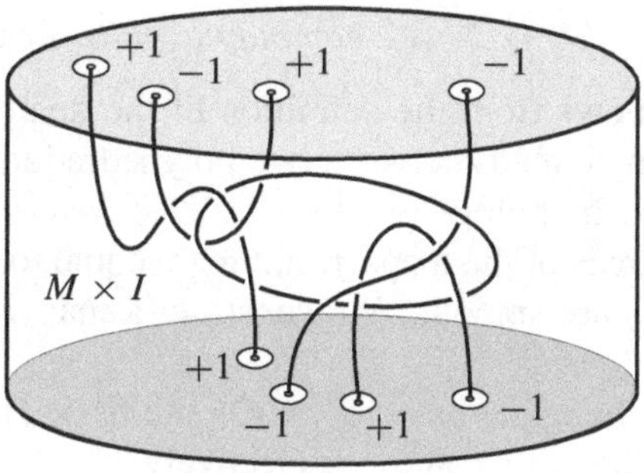


Figure 13. The sum of the numbers on the upper base of cylinder $M \times I$ equals the sum of the numbers on the lower base.

## 1.8 Applications of the degree of a map

In this section we consider several classical examples of applications of the degree of maps as well as one new example of such an application. Let us start with the homotopic classification of maps from the circle to itself. Each map $f: S_1^1 \to S_2^1$ from one circle to another can be viewed as a parameterized path in $S_2^1$, which starts and ends at the same point. Moving along this path we perform an integer number of windings around the circle. This number coincides with the degree of $f$, or, rather, the notion of the degree is a formalization of the intuitively clear but not very rigorous notion of "the total number of windings". Since the annulus is homotopy equivalent to the circle, we can consider closed paths in the annulus instead of in the circle.

**Exercise 37.** What is the total number of windings around the marked point that is performed by the closed path shown in Figure 14?

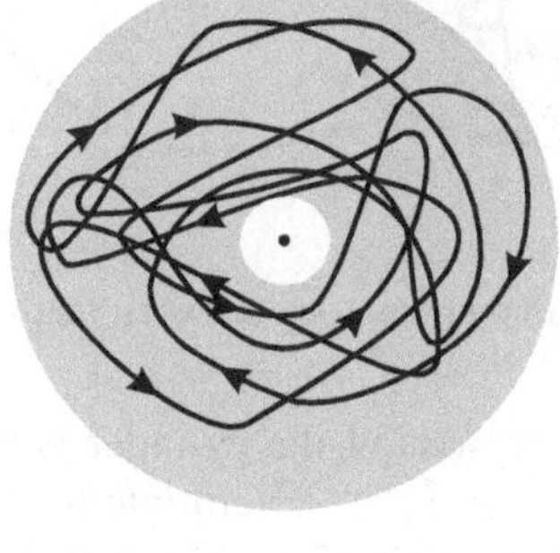

Figure 14. A closed path in an annulus determines a map from a circle to a circle.

**Theorem 10.** *Maps* $f, g\colon S^1 \to S^1$ *are homotopic* $\Longleftrightarrow$ $\deg f = \deg g$.

*Proof.* The part $\Longrightarrow$ follows from the definition of the degree and from Theorem 8, which states that homotopic maps between two polyhedra induce identical homomorphisms of their respective homology groups.

Suppose that the degrees of the maps $f$ and $g$ are equal to the same number $n$. Let us assume that the maps are smooth. We construct a map $F\colon S^1 \times I \to S^1$ in the following way.

1. On the lower and the upper bases $S^1 \times \{0\}$ and $S^1 \times \{1\}$ of the cylinder $S^1 \times I$ the map $F$ should coincide with $f$ and $g$ respectively.

2. Let $y_0 \in S^1$ be a common regular value of the maps $f$ and $g$ such that its full pre-images $x_1, \dots, x_k \in S^1 \times \{0\}$ and $x'_1, \dots, x'_m \in S^1 \times \{1\}$, situated at the bases of the cylinder, are non-empty. As above, these points are marked by certain numbers $\pm 1$ such that the numbers attributed to the points of the lower base, as well as those attributed to the points of the upper base, sum up to $n$. Let us join these points by disjoint arcs inside the cylinder in such a way that each arc whose ends are on the same base would join points with opposite signs and that each arc whose ends are on distinct bases would join points with identical signs. Obviously, at least one arc of the latter type should be present.

3. We map the arcs thus chosen to the point $y_0$ and then extend this map to a map from some ribbon neighbourhoods of the arcs to a small arc $\ell_0 \subset S^1$, as shown in Figure 15.

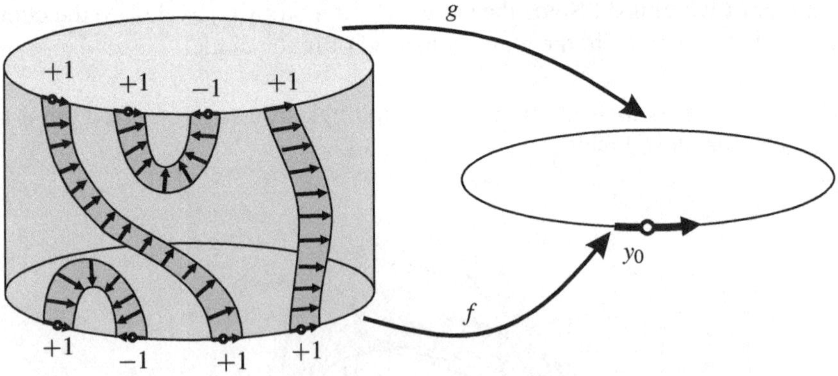


Figure 15. Constructing a homotopy between maps of the same degree.

4. Each of the remaining regions of the cylinder is homeomorphic to a disc, and its boundary is mapped to the arc $\ell_1 \subset S^1$ complementary to the arc $\ell_0$. Therefore the obtained map defined on the union of the bases of the cylinder and of the ribbons can be extended to a map from the whole cylinder to $S^1$. This yields the desired homotopy. □

The above theorem implies that each map from a circle to itself is homotopic either to the constant map to a point (when the degree is equal to zero) or to the $n$-fold winding in the positive or in the negative direction, depending on the sign of the degree.

Another application of the degree of a map is Whitney's theorem on the regular homotopy classification of immersions of the circle into the plane.

**Definition.** A smooth map $f: S^1 \to \mathbb{R}^2$ is called an *immersion* if it does not have any singular points, *i.e.* if the tangent vector never vanishes. Such maps are often called *regular curves*.

**Definition.** Immersions $f, g: S^1 \to \mathbb{R}^2$ are called *regular homotopic* if there exists a smooth map $F: S^1 \times I \to \mathbb{R}^2$ such that $F(x, 0) = f(x)$ and $F(x, 1) = g(x)$ for all $x$ and the map $F_t: S^1 \to \mathbb{R}^2$ determined by the formula $F_t(x) = F(x, t)$ is an immersion for all $t \in I$.

It is relevant to recall here that any smooth map $F$ is infinitely differentiable. Hence the tangent vector of $F_t$ depends continuously (and even smoothly) both on the parameter $t$ and on the initial point. This implies that the homotopy shown in Figure 16 is not regular, although the curve looks like an immersion at each moment $t$. The reason is that the tangent vector at the marked point vanishes.

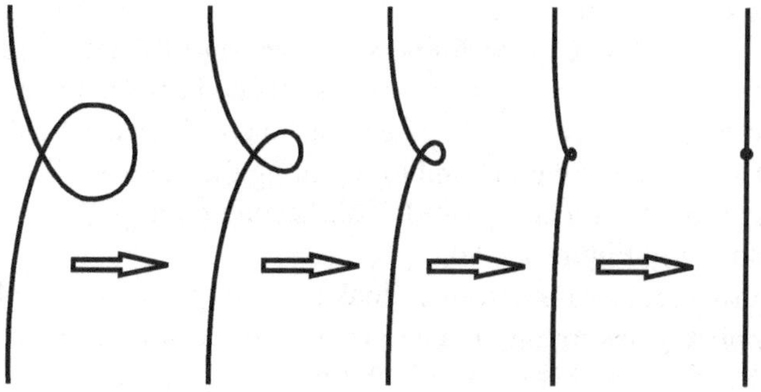

Figure 16. Tightening of a loop.

Let $f: S^1 \to \mathbb{R}^2$ be an immersion. Assign to each point $x \in S^1$ the endpoint of the unit vector $v(t)/|v(t)|$ with origin at the center of the coordinate system, where $v(t) = f'(t)$ is the tangent vector. We obtain a map $v_f: S^1 \to S^1$. It turns out that the regular homotopy class of the immersion $f$ is fully determined by the degree of the map $v_f$. This number is called the *writhe number* of $f$ and is denoted by $w(f)$.

**Theorem 11** (Whitney's Theorem). *For any integer $n$ there exists an immersion $f: S^1 \to \mathbb{R}^2$ such that $w(f) = n$. Furthermore, two immersions $f, g: S^1 \to \mathbb{R}^2$ are regular homotopic $\iff w(f) = w(g)$.*

*Proof.* The validity of the first claim of the theorem follows from Figure 17. The figure shows certain immersions for every writhe number $n \geqslant 0$. Let us call them *standard.* Standard immersions with negative writhe numbers are obtained from those with positive writhe numbers by reversing the direction of winding. If $n \neq 0$, the

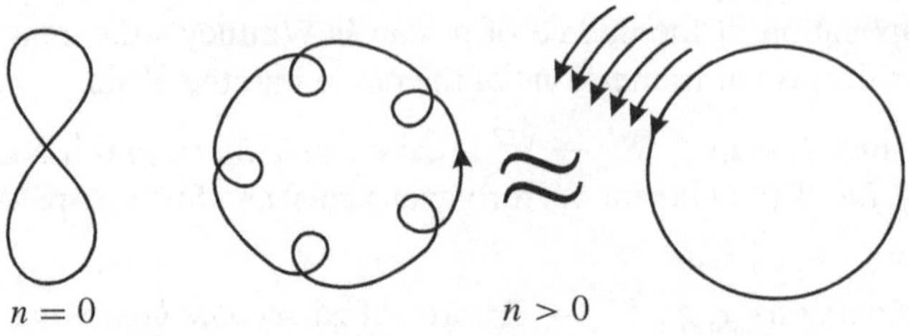


Figure 17. Standard immersions of the circle.

absolute value of $n$ is the number of loops plus 1, and its sign is determined by the direction in which we go around the curve. The figure also shows another variety of standard immersions for $n \neq 0$: it is just the $n$-fold winding around the standard circle in the desired direction. Since under a regular homotopy of an immersion $f$ the map $v_f$ is subjected to a homotopy, the part $\Longrightarrow$ of the second claim of the theorem follows from Theorem 10. To prove the implication $\Longleftarrow$ we show that every immersion is regular homotopic to a standard one.

1. Let $f\colon S^1 \to \mathbb{R}^2$ be a given immersion. We start by detecting there a *simple loop, i.e.* a piece of the immersed circle which starts and ends at the same point of the plane and has no other points of self-intersection. Such a loop can be easily found. We should just start at an arbitrary point and move along the curve until we encounter one of the points that we have already passed. This second point will be the starting point of the desired loop. See Figure 18, left.

2. Now we use a regular homotopy to make this loop very small, almost invisible, but without completely destroying it. Then we may disregard this loop henceforth and operate with the curve as if this loop were not present. See Figure 18, right.

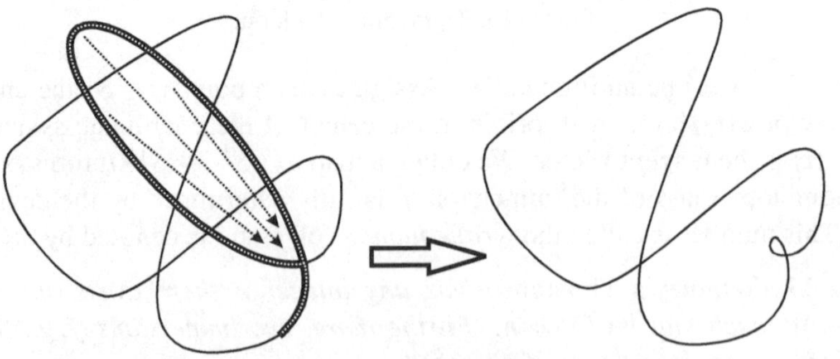

Figure 18. Taming of the loops.

3. Let us perform such operations for as long as possible. In the end we get a circle with several small loops. These loops are divided into interior and exterior ones. Figure 19 shows a regular homotopy destroying a pair of neighbouring scalene loops. This trick allows us to ensure that all loops are placed from one side only.

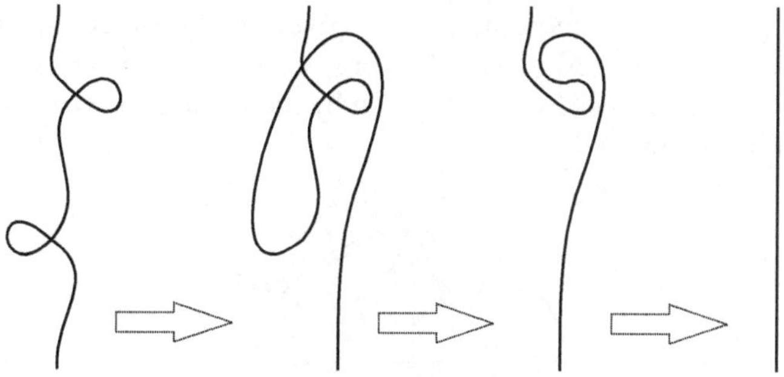

Figure 19. Destroying neighbouring loops.

4. If all the loops are interior ones, then we have a standard immersion. Assume that all the loops are exterior and that their number is greater than 1. Choose one of these loops and move all the other loops inside it by the regular homotopy shown in Figure 20. Then we destroy the obtained pair of neighbouring scalene loops and again get a standard immersion.

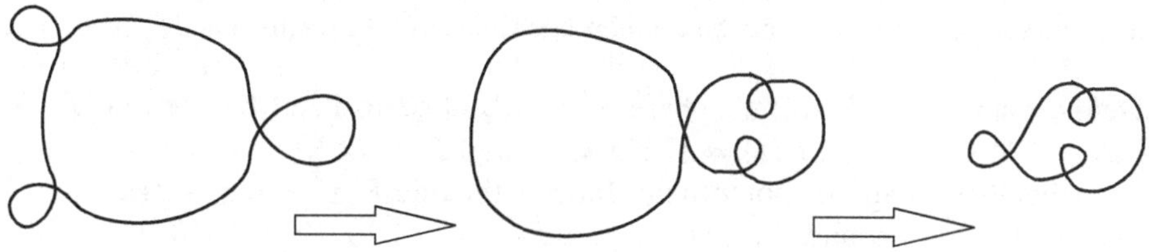

Figure 20. Turning exterior loops into interior ones.

5. In the single remaining case there is only one exterior loop and no interior ones, hence we have a standard immersion with writhe number 0 (eight-figure). □

**Exercise 38.** Give a regular homotopy classification of immersions of the circle into $S^2$.

Let us prove now the Fundamental Theorem of Algebra, which states that every polynomial equation over the complex field is solvable. Of course, there are many nice proofs of this theorem but the one described below is perhaps the most conceptual.

**Theorem 12** (Fundamental Theorem of Algebra). *Any polynomial of non-zero degree has a root (complex or real).*

*Proof.* Let $P(z) = z^n + a_{n-1}z^{n-1} + \cdots + a_0$ be the given polynomial. Present the sphere $S^2$ as the extended complex plane $\mathbb{C} \cup \infty$. Such presentation endows the sphere with the structure of a complex manifold. It is convenient to cover it with two charts, the complement of the south pole and the complement of the north pole. The homeomorphisms from these regions to the complex plane $\mathbb{C}_z$ with coordinate $z$ and the complex plane $\mathbb{C}_w$ with coordinate $w$ are given by the stereographic projections $\varphi\colon S^2 \setminus S \to \mathbb{C}_z$ and $\psi\colon S^2 \setminus N \to \mathbb{C}_w$. See Figure 21, which explains why the change of charts $\psi\varphi^{-1}\colon \mathbb{C}_z \setminus \{0\} \to \mathbb{C}_w \setminus \{0\}$ is given by the formula $w = 1/z$ (provided that the diameter of the sphere is 1).

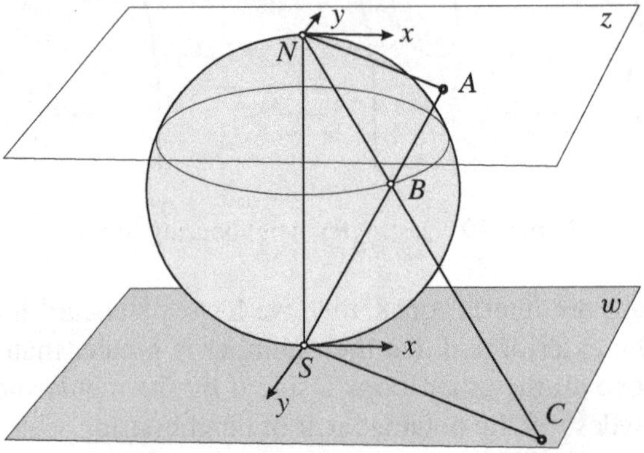


Figure 21. The similarity of triangles $SNC$ and $SAN$ implies that $|z||w| = 4R^2$, where $R$ is the radius of the sphere and $z$ and $w$ are the complex coordinates of the points $A$ and $C$, respectively.

Define maps $P, E$ from the sphere $S^2 = \mathbb{C}_z \cup \infty$ to itself by the rule $P(z) = z^n + a_{n-1}z^{n-1} + \cdots + a_0$, $E(z) = z^n$ if $z \neq \infty$ and $P(z) = E(z) = \infty$ if $z = \infty$. Let us prove that these maps are homotopic. Indeed, the rule $F(z,t) = z^n + t(a_{n-1}z^{n-1} + \cdots + a_0)$ for $z \neq \infty$ and $F(z,t) = \infty$ for $z = \infty$, for all $t \in [0,1]$, defines a homotopy between them, since $F(z,1) = P(z)$ and $F(z,0) = E(z)$. At first glance, this homotopy may have a nasty behaviour in the neighbourhood of the point $z = \infty$ (for instance, be discontinuous). However, this is easily seen not to be the case by passing to the coordinate $w = 1/z$. The homotopy formula written with respect to the coordinate $w$ has the form,

$$G(w,t) = \frac{w^n}{1 + t(a_{n-1}w^{n-1} + \cdots + a_0)},$$

therefore for small $w$ (when $|a_{n-1}w^{n-1} + \cdots + a_0| < 1$ and so the denominator is never zero) the homotopy is not only continuous but also differentiable. For the same reason the maps $P$ and $E$ are not only continuous but also holomorphic, *i.e.* differentiable in the complex sense.

Let us now compare the following facts.

1. The degree of $E$ is equal to $n$. Indeed, the pre-image of any point $a \neq 0, \infty$ of $S^2$ consists of exactly $n$ points ($n$-th degree roots of $a$), and at each of them the Jacobian is positive. The latter holds for any regular point of any holomorphic function $f : \mathbb{C} \to \mathbb{C}$. Indeed, if this function is written in real coordinates as $f(x + iy) = u(x, y) + v(x, y)i$, then by the Cauchy–Riemann conditions $u_x = v_y, u_y = -v_x$ (for brevity, we use this notation for the partial derivatives) the Jacobian $\left| \begin{smallmatrix} u_x & u_y \\ v_x & v_y \end{smallmatrix} \right| = u_x^2 + u_y^2$ is indeed always positive.

2. Since the map $P$ is homotopic to the map $E$, the degree of $P$ is also equal to $n$.

3. Finally, if the polyhedron $P(z)$ does not have a root, then the degree of the map $P$ determined by it is equal to 0, since the pre-image of 0 (which is a regular value) is empty.

Together these three facts imply that any polynomial of non-zero degree has a root. □

It is useful to add that in fact we have proven much more: for almost all right-hand terms $a \in \mathbb{C}$ the equation $P(z) = a$ has exactly $n$ roots. More precisely, the number of distinct roots is $n$ for any regular value $a$ of the map $P$.

As one more application of the degree, let us prove the Hairy Ball Theorem. It claims that a hairy ball cannot be combed such that all hairs lay flat.

**Theorem 13.** *Any continuous tangent vector field on $S^2$ has a singular point, i.e. a point where it vanishes. (See Figure 22, left, which shows the trajectories of a vector field with just one singular point.)*

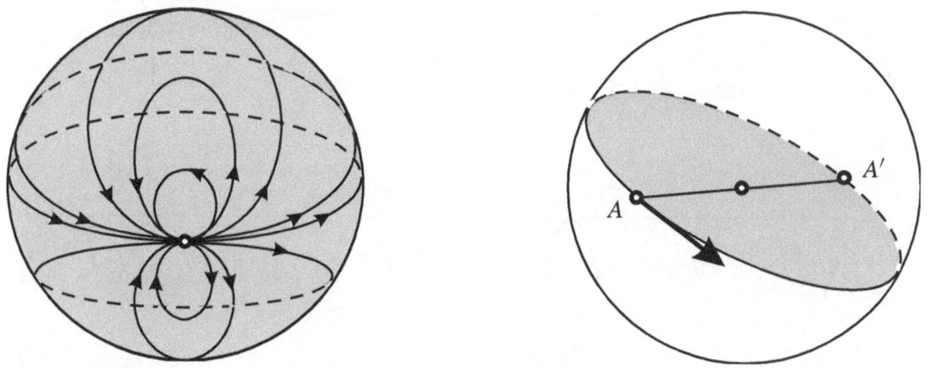


Figure 22. Trajectories of a vector field with one singular point (left). Each point $A$ moves towards its opposite $A'$ along a big semi-circle (right).

*Proof.* Denote by $r$ the antipodal map of the sphere to itself, which takes each point $A \in S^2$ to its diametrically opposite point $A'$. The degree of this map is equal to $-1$, hence it is not homotopic to the identity map, whose degree is 1. The proof of the theorem is by contradiction. Assume that some given vector field does not have any singular points. Let us force each point $A \in S^2$ to move gradually towards its antipode $A'$ along a semicircle of the big circle that passes through the points $A$, $A'$ and touches the vector coming out of $A$. In fact, out of the two semi-circles we choose the one towards which the vector is pointed. See Figure 22, right. It is easy to show that this rule defines a homotopy between the antipodal map $r$ and the identity, which is impossible. □

Now we describe a recent example of an application of the degree, a solution of the question about the number of common tangent lines of two immersed circles. The author is grateful to M. Polyak, who introduced him to this problem and its beautiful solution.

Let $f, g \colon S^1 \to \mathbb{R}^2$ be two immersed circles, whose images (which are immersed closed curves) lie in the interiors of two complementary half-planes. How many common tangent lines can they have? If the curves are standard round circles then the answer is evident: exactly 4. But how many common tangent lines are there in a general case, when, let us say, the writhe numbers $w(f)$, $w(g)$ are taken arbitrarily? The question in this form does not make much sense, since a small perturbation of one of the curves may lead to the appearance of one or more common tangent lines, see Figure 23. However, the statement of the problem can be modified a bit to make it meaningful.

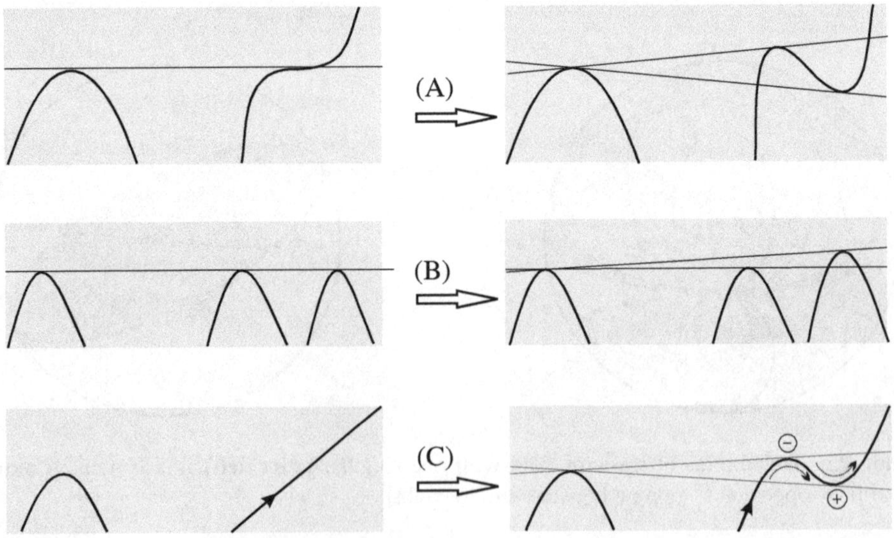


Figure 23. Three reasons for the appearance or disappearance of common tangent lines: cubic tangency (A), double tangency (B), zigzag (C).

The reason for the emergence of new tangents in the case (A) consists in the presence of a degenerate (in our case cubic) tangency. To avoid it we require that all common tangents be non-degenerate. This means that the curvatures of both curves in all points of tangency must be distinct from zero, *i.e.* that the tangency must be quadratic.

In the case (B) the obstacle is removed by taking into account the multiplicity of tangents. Namely, if a line touches the first curve in points $a_1, \dots, a_k$ and it touches the second curve in points $b_1, \dots, b_l$, then it should be considered as $kl$ tangents, one for each pair of points $a_i$, $b_j$. This is quite natural, since exactly this many tangents appear under a small perturbation of the curves.

The case (C) is more difficult, since there we cannot get rid of the new tangents. The solution is to endow the common tangents with signs $\pm 1$ in such a way that the tangents from each arising pair have distinct signs and hence cancel each other out under counting. Let $l$ be a line which touches curves $f$ and $g$ at points $f(t_0)$ and $g(\tau_0)$ respectively, where $t$ and $\tau$ are the parameters on the two copies of $S^1$. Assign to the point $f(t_0)$ a sign $\delta_1 = \pm 1$ depending on the direction of the turn (positive or negative) of the velocity vector $v(t) = f'(t)$ when the parameter $t$ passes through the value $t = t_0$. The sign $\delta_2$ of the point $g(\tau_0)$ is defined similarly. Now we are ready to define the sign $\varepsilon_l$ of the common tangent $l$: it is equal to the product $\delta_1\delta_2$. It is easy to check that the common tangents arising in the case (C) have opposite signs.

**Definition.** Suppose that all the common tangents of two immersed curves $f$, $g$ are non-degenerate, and $n_+$ of them are positive, $n_-$ are negative. Then the difference $n_+ - n_-$ is called the *reduced number* of their common tangents and is denoted by $t(f, g)$.

**Theorem 14.** *Let $f, g$ be two immersions of the circle to the plane such that their images lie inside two complementary semi-planes and all their common tangent lines are non-degenerate. Then $t(f, g) = 4w(f)w(g)$, where $w(f)$, $w(g)$ are the writhe numbers of the immersions $f, g$.*

*Proof.* Let $f : S^1_t \to \mathbb{R}^2$, $g : S^1_\tau \to \mathbb{R}^2$ be the given immersions, where $t$ and $\tau$ are the parameters of the respective circles. Denote by $u(t, \tau)$ the vector going from the point $f(t)$ to the point $g(\tau)$. Denote also by $\alpha(t, \tau)$, $\beta(t, \tau)$ the angles between $u(t, \tau)$ and the velocity vectors $v_f(t) = f'(t)$, $v_g(\tau) = g'(\tau)$, measured in the direction of the positive turn, see Figure 24. The angles are of course considered modulo $2\pi$. Then assigning to the pair $(t, \tau)$ the pair $(\alpha(t, \tau), \beta(t, \tau))$ defines a map $C_{fg} : S^1_t \times S^1_\tau \to S^1_\alpha \times S^1_\beta$ between two tori.

We claim that the reduced number $t(f, g)$ of the common tangents is equal to the degree of this map multiplied by 4, *i.e.* to the number $4\deg(C_{fg})$. The proof of this claim is given by the following observations.

1. The line passing through points $f(t)$, $g(\tau)$ touches the curves at these points if and only if the angles $\alpha(t, \tau)$, $\beta(t, \tau)$ are 0 modulo $\pi$, *i.e.* when the point $(t, \tau)$ belongs to the pre-image of one of the points $(0, 0)$, $(0, \pi)$, $(\pi, 0)$, $(\pi, \pi)$.

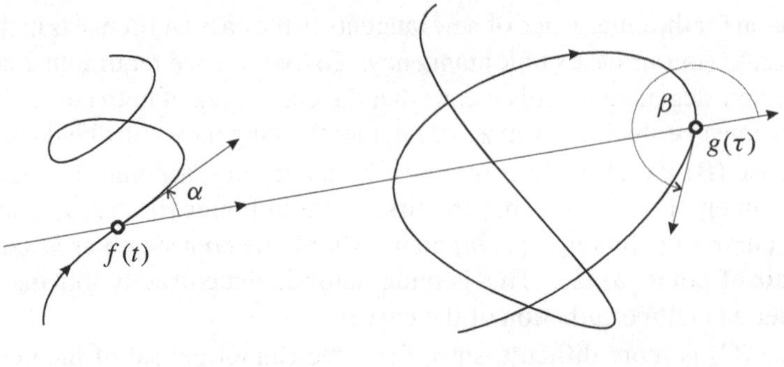


Figure 24. Assignment to the pair $(t, \tau)$ of the pair $(\alpha, \beta)$.

2. The points $(0, 0)$, $(0, \pi)$, $(\pi, 0)$, $(\pi, \pi)$ are regular values of the map $C_{fg}$. This follows from the non-degeneracy of all common tangents, since if some line $l$ touches the first curve at a point $f(t)$ and the second curve at a point $g(\tau)$, then $\alpha'_t(t, \tau) \neq 0$, $\beta'_\tau(t, \tau) \neq 0$, while $\alpha'_\tau(t, \tau) = \beta'_t(t, \tau) = 0$. Therefore the determinant $\alpha'_t(t, \tau)\beta'_\tau(t, \tau)$ of the Jacobi matrix is non-zero. Moreover, since the signs of the derivatives $\alpha'_t(t, \tau)$, $\beta'_\tau(t, \tau)$ show the directions of the turns of the tangent vectors, the sign of this determinant coincides with the sign of $l$.

3. It follows that the difference $m_+ - m_-$ between the number of positive tangents and that of negative tangents of type $(0, 0)$ coincides with the degree of the map $\deg(C_{fg})$. Similar statements are true for tangents of the types $(0, \pi)$, $(\pi, 0)$, $(\pi, \pi)$. Therefore $t(f, g) = 4 \deg(C_{fg})$.

Notice that when $f, g$ are replaced by curves regular homotopic to them (the intermediate curves of the regular homotopy should not intersect), the map $C_{gf}$ is also replaced by a homotopic one. Hence its degree, and therefore the reduced number of tangents $t(f, g)$, are preserved. Hence to complete the proof it suffices to show that the statement is true for all standard immersions of the circle. This is easy. For instance, in the case when $w(f) \neq 0$, $w(g) \neq 0$ the standard immersions can be taken to be suitable windings around non-intersecting standard circles. Then each of the four common tangents to these circles gives rise to $\deg(C_{gf})$ tangents to the windings. □

## 1.9 Relative homology

It is possible to assign a certain sequence of homology groups not only to a topological space but also to a pair of spaces, one of which is a subspace of the other one. The homology groups of pairs of spaces (*relative homology*) are closely tied with those of each of the two individual spaces and hence relate them to each other. The definition and properties of the relative homology groups do not differ much from those of the absolute homology groups. Therefore we restrict our discussion of them to a few brief comments. It is also useful to keep in mind that the relative homology groups of a complex (or a space) with respect to the empty set coincide with the absolute ones, *i.e.* the groups $H_n(K)$ and $H_n(K, \emptyset)$ are always isomorphic.

Let $L$ be a subcomplex of some complex $K$. The *relative chain group* $C_n(K, L)$ of the pair $(K, L)$ is defined as the free Abelian group freely generated by all simplices with interiors in $K \setminus L$. The operator $\partial_n : C_n(K, L) \to C_{n-1}(K, L)$ is defined by the formula

$$\partial_n(\sigma) = \sum_{\delta_i} \varepsilon_i \delta_i,$$

where the numbers $\varepsilon_i$ have the same meaning as in the absolute case and the sum is taken over all those $(n-1)$-dimensional simplices $\delta_i$ of $K$ whose interiors do not intersect $L$.

It is checked directly that the relative chain groups and the operators $\partial_n$ form a chain complex $C(K, L)$, which is called the *relative chain complex* of the pair $(K, L)$. Its homology groups are called the *relative homology groups* of the pair $(K, L)$. To provide more details, the group of cycles $A_n(K, L)$ is by definition the kernel of the homomorphism $\partial_n$, the group of boundaries $B_n(K, L)$ is the image of $\partial_{n+1}$, and the relative homology group $H_n(K, L)$ is the quotient group $A_n(K, L)/B_n(K, L)$. As in the absolute case, any simplicial map of pairs $f : (K, L) \to (K_1, L_1)$ (*i.e.* a simplicial map $f : K \to K_1$ such that $f(L) \subset L_1$) induces homomorphisms $f_* : H_n(K, L) \to H_n(K_1, L_1)$.

**Theorem 15.** *Suppose that complexes $X$, $Y$ intersect along a complex $Z$, which is a subcomplex in both $X$ and $Y$. Then the embedding $i$ of the pair $(X, Z)$ to the pair $(X \cup Y, Y)$ induces isomorphisms of the relative homology groups.*

*Proof.* The theorem is evident since the chain complexes of the pairs $(X, Z)$, $(X \cup Y, Y)$ coincide (being constructed over the same set of simplices). □

In order to better understand the notion of relative homology it is recommended to think over the question, what is the difference between the relative homology groups of a pair $(K, L)$ and the absolute homology groups of the closure of the difference $K \setminus L$?

## 1.10 The exact homology sequence

The relation between the absolute homology groups of individual spaces and the relative homology groups of pairs can be described very well using the language of exact sequences.

**Definition.** A sequence of groups and their homomorphisms

$$\cdots \longrightarrow A_{n+1} \xrightarrow{\varphi_{n+1}} A_n \xrightarrow{\varphi_n} A_{n-1} \longrightarrow \cdots$$

is called *exact* if the kernel of each subsequent homomorphism coincides with the image of the preceding one, *i.e.* if $\operatorname{Ker} \varphi_n = \operatorname{Im} \varphi_{n+1}$ for all $n$.

This definition implies that any exact sequence of groups infinite in both directions is a chain complex. All the homology groups of that complex are trivial. The reverse is also true: any chain complex with trivial homology groups is an exact sequence. Therefore the homology groups of a chain complex provide a *measure of its inexactness* in some sense.

**Exercise 39.** What can one say about the homomorphism $\varphi$ in the exact sequence $0 \longrightarrow A \xrightarrow{\varphi} B$? In the exact sequence $A \xrightarrow{\varphi} B \longrightarrow 0$? In the exact sequence $0 \longrightarrow A \xrightarrow{\varphi} B \longrightarrow 0$?

**Exercise 40.** Prove that if the sequence $0 \longrightarrow A \longrightarrow B \longrightarrow C \longrightarrow 0$ is exact, then $C = B/A$.

Exact sequences of chain complexes and chain maps are defined similarly to those of groups: it is required that the groups of each dimension and homomorphisms between them form an exact sequence. In particular, a short sequence

$$0 \longrightarrow C' \xrightarrow{i} C \xrightarrow{p} C'' \longrightarrow 0$$

of chain complexes is exact if and only if for each $n$ the short sequence

$$0 \longrightarrow C'_n \xrightarrow{i} C_n \xrightarrow{p} C''_n \longrightarrow 0$$

of their chain groups is exact.

Examples of short exact sequences of chain complexes are easiest to obtain via geometric constructions. The definition of the relative chain complex $C(K, L)$ of a pair $(K, L)$ of simplicial complexes immediately implies that the sequence

$$0 \longrightarrow C(L) \xrightarrow{i} C(K) \xrightarrow{p} C(K, L) \longrightarrow 0,$$

with the homomorphism $i$ induced by the embedding of $L$ into $K$ and the homomorphism $p$ obtained by forgetting all those simplices of $K$ that are contained in $L$, is exact.

**Theorem 16.** *Suppose that chain complexes $C$, $C'$, $C''$ are related by the short exact sequence*

$$0 \longrightarrow C' \xrightarrow{i} C \xrightarrow{p} C'' \longrightarrow 0.$$

*Then it is possible to define certain homomorphisms $\delta$ in such a way that the long sequence of homology groups*

$$\cdots \longrightarrow H_n(C') \xrightarrow{i_*} H_n(C) \xrightarrow{p_*} H_n(C'') \xrightarrow{\delta} H_{n-1}(C') \longrightarrow \cdots$$

*is exact.*

*Proof.* To define the desired homomorphisms $\delta$ we draw the following commutative diagram and apply to it the method of diagrammatic search.

$$\begin{array}{ccccccccc}
 & & \vdots & & \vdots & & \vdots & & \\
 & & \downarrow{\scriptstyle\partial} & & \downarrow{\scriptstyle\partial} & & \downarrow{\scriptstyle\partial} & & \\
0 & \longrightarrow & C'_{n+1} & \xrightarrow{i} & C_{n+1} & \xrightarrow{p} & C''_{n+1} & \longrightarrow & 0 \\
 & & \downarrow{\scriptstyle\partial} & & \downarrow{\scriptstyle\partial} & & \downarrow{\scriptstyle\partial} & & \\
0 & \longrightarrow & C'_{n} & \xrightarrow{i} & C_{n} & \xrightarrow{p} & C''_{n} & \longrightarrow & 0 \\
 & & \downarrow{\scriptstyle\partial} & & \downarrow{\scriptstyle\partial} & & \downarrow{\scriptstyle\partial} & & \\
0 & \longrightarrow & C'_{n-1} & \xrightarrow{i} & C_{n-1} & \xrightarrow{p} & C''_{n-1} & \longrightarrow & 0 \\
 & & \downarrow{\scriptstyle\partial} & & \downarrow{\scriptstyle\partial} & & \downarrow{\scriptstyle\partial} & & \\
 & & \vdots & & \vdots & & \vdots & &
\end{array}$$

In terms of cycles the homomorphism $\delta$ is given by the formula $\delta = i^{-1}\partial p^{-1}$. This formula should be understood as follows. Take $h \in H_n(C'')$. Choose a cycle $a \in A_n(C'') \subset C''_n$ representing $h$. Since the homomorphism $p$ is surjective (notice that the subsequent term of the exact sequence is 0!), there exists $t \in p^{-1}(a) \in C_n$. The commutativity of our diagram implies that $p\partial(t) = \partial p(t) = \partial a = 0$, therefore $\partial t \in \operatorname{Ker} p = \operatorname{Im} i$. Thus, there is a well-defined element $x = i^{-1}\partial(t)$. Since $i\partial(x) = \partial i(x) = \partial^2(t) = 0$ and $i$ is injective, we have $\partial(x) = 0$, *i.e.* $x$ is a cycle and therefore defines a certain element $\bar{h} \in H_{n-1}(C')$. We set $\delta(h) = \bar{h}$.

The further proof is decomposed into several steps, stated below as a string of exercises. The exercises can also be solved via the method of diagrammatic search. □

**Exercise 41.** Prove that the element $\bar{h}$ does not depend on the choice of $t \in p^{-1}(a)$.

**Exercise 42.** Prove that the element $\bar{h}$ does not depend on the choice of the cycle $a$ representing $h$.

**Exercise 43.** Prove that the sequence of homology groups thus obtained is exact.

It should be noticed that the homomorphisms defined in the proof of Theorem 16 are *functorial* in the following sense. Let $\bar{\varphi} = \{\varphi', \varphi, \varphi''\}$ be a map

$$\begin{array}{ccccccccc} 0 & \longrightarrow & C' & \xrightarrow{i} & C & \xrightarrow{p} & C'' & \longrightarrow & 0 \\ & & \downarrow{\scriptstyle\varphi'} & & \downarrow{\scriptstyle\varphi} & & \downarrow{\scriptstyle\varphi''} & & \\ 0 & \longrightarrow & D' & \xrightarrow{i'} & D & \xrightarrow{p'} & D'' & \longrightarrow & 0 \end{array}$$

between two short sequences of chain complexes such that all the squares are commutative. Then $\varphi'_*\delta = \delta\varphi''_*$ for all $n$, where the homomorphisms $\varphi''_*\colon H_n(C'') \to H_n(D'')$ and $\varphi'_*\colon H_{n-1}(C') \to H_{n-1}(D')$ are induced by the homomorphisms $\varphi''$, $\varphi'$. This fact is also established by means of a diagrammatic search.

**Theorem 17.** *For any pair $(K, L)$ of simplicial complexes the homology groups of these complexes and the relative homology groups of the pair $(K, L)$ form the long exact sequence*

$$\cdots \longrightarrow H_n(L) \xrightarrow{i_*} H_n(K) \xrightarrow{p_*} H_n(K, L) \xrightarrow{\delta} H_{n-1}(L) \longrightarrow \cdots.$$

*Proof.* Since the chain complexes $C(L)$, $C(K)$, $C(K, L)$ form a short exact sequence, Theorem 17 is an immediate corollary of Theorem 16. □

**Exercise 44.** Determine all the homology groups of the $n$-dimensional sphere.

Another corollary of Theorem 16 is the exact Mayer–Vietoris sequence. Let $K_1$, $K_2$ be subcomplexes of a simplicial complex $K$ such that $K = K_1 \cup K_2$.

For $m = 1, 2$ denote by $i_m$, $j_m$ the homomorphisms $C(K_1 \cap K_2) \to C(K_m)$ and $C(K_m) \to C(K)$ induced by the natural embeddings $K_1 \cap K_2 \to K_m$ and $K_m \to K$ respectively. Then we can write the short sequence of chain complexes

$$0 \longrightarrow C(K_1 \cap K_2) \xrightarrow{i} C(K_1) \oplus C(K_2) \xrightarrow{j} C(K_1 \cup K_2) \longrightarrow 0,$$

where $i = i_1 \oplus (-i_2)$ and $j = j_1 + j_2$. In other words, $i(x) = (i_1(x), -i_2(x))$ for any chain $x \in C_n(K_1 \cap K_2)$ and $j(y_1, y_2) = j_1(y_1) + j_2(y_2)$ for any chain $(y_1, y_2) \in C_n(K_1) \oplus C_n(K_2)$.

**Exercise 45.** Show that this sequence is exact.

**Theorem 18.** *Let $K_1, K_2$ be subcomplexes of a simplicial complex $K$ such that $K = K_1 \cup K_2$. Then there exist functorial homomorphisms $\delta\colon H_n(K_1 \cup K_2) \to H_{n-1}(K_1 \cap K_2)$ such that the homology groups of the complexes $L = K_1 \cap K_2$, $K_1$, $K_2$, and $K$ form the long exact sequence*

$$\cdots \longrightarrow H_n(L) \xrightarrow{i_*} H_n(K_1) \oplus H_n(K_2) \xrightarrow{j_*} H_n(K) \xrightarrow{\delta} H_{n-1}(L) \longrightarrow \cdots.$$

*Proof.* The proof follows from Theorem 16. □

## 1.11 Axiomatic point of view on homology

An axiomatic description of the notion of homology allows us to establish the equivalence of different homology theories with relative ease. Each homology theory is a functor $\mathcal{H}$ from the category of pairs of polyhedra to the category of sequences of Abelian groups. The groups assigned to a pair of polyhedra $(P, Q)$ are called the homology groups of this pair and are denoted by $\mathcal{H}_n(P, Q)$. Let us list the axioms which should be satisfied by any homology functor $\mathcal{H}$.

**I. Homotopy axiom**. The functor $\mathcal{H}$ must be *homotopic*. This means that homotopic maps between polyhedra should induce identical homomorphisms of the homology groups.

**II. Long exact sequence of a pair axiom**. The functor $\mathcal{H}$ must be exact. This means that for each pair of polyhedra $(P, Q)$ there must be a long exact sequence corresponding to it, namely the following one:

$$\cdots \longrightarrow \mathcal{H}_n(Q) \xrightarrow{i_*} \mathcal{H}_n(P) \xrightarrow{p_*} \mathcal{H}_n(P, Q) \xrightarrow{\delta} \mathcal{H}_{n-1}(Q) \longrightarrow \cdots,$$

where the homomorphisms $i_*$ and $p_*$ are induced by the embeddings of $Q$ to $P$ and of the pair $(P, \phi)$ to $(P, Q)$, and the homomorphisms $\delta$ are functorial with respect to maps of pairs.

**III. Excision axiom**. Suppose that polyhedra $X$, $Y$ intersect along a subpolyhedron $Z$. Then the embedding $i$ of the pair $(X, Z)$ to the pair $(X \cup Y, Y)$ must induce isomorphisms of the respective homology groups.

**IV. Dimension axiom**. The homology groups of the point must be the following ones:

$$\mathcal{H}_n(*) = \begin{cases} 0, & n \neq 0, \\ \mathbb{Z}, & n = 0. \end{cases}$$

**Theorem 19.** *The simplicial homology functor $H$ satisfies axioms* I–IV.

*Proof.* The proof follows from Theorems 8, 17, 15, and 5. □

In order to be able to compare functors, recall the definitions of the natural transformation of functors and of the equivalence of functors. Let $F_1$, $F_2$ be functors from a category $\mathsf{G}_1$ to a category $\mathsf{G}_2$. Suppose that to each object $A$ of the category $\mathsf{G}_1$ we have assigned a morphism $\varphi_A$ from the object $F_1(A)$ to the object $F_2(A)$ in such a way that for any morphism $f\colon A \to B$ in $\mathsf{G}_1$ there is the equality $\varphi_B F_1(f) = F_2(f)\varphi_A$, *i.e.* the following diagram is commutative.

$$\begin{array}{ccc} F_1(A) & \xrightarrow{F_1(f)} & F_1(B) \\ \downarrow{\scriptstyle \varphi_A} & & \downarrow{\scriptstyle \varphi_B} \\ F_2(A) & \xrightarrow{F_2(f)} & F_2(B) \end{array}$$

In this case it is said that there is a *natural transformation* of the functor $F_1$ into the functor $F_2$. If for any object $A$ in $\mathsf{G}_1$ the morphisms $\varphi_A$ are isomorphisms, then it is said that the functors are *equivalent*.

Now we are ready to state the uniqueness theorem.

**Theorem 20.** *Suppose that $\mathcal{H}$ is an exact homotopic functor from the category of pairs of polyhedra to the category of sequences of Abelian groups that satisfies the excision axiom and the dimension axiom. Suppose that there exists a natural transformation of the simplicial homology functor $H$ to the functor $\mathcal{H}$ such that the morphism $\varphi_*$ corresponding to the point is an isomorphism. Then these functors are equivalent. In particular, for any pair $(P, Q)$ of compact polyhedra the groups $\mathcal{H}_i(P, Q)$ are isomorphic to the respective groups $H_i(P, Q)$.*

To ensure a better understanding let us describe, very roughly, a scheme of the proof of the uniqueness theorem. By the dimension axiom the groups $H_n(*)$ and $\mathcal{H}_n(*)$ coincide. A simultaneous consideration of the exact sequences of the pairs $(D^n, S^{n-1})$ and $(S^{n-1}, D^{n-1})$, where $D^n$ is the $n$-dimensional ball, $S^{n-1}$ is the $(n-1)$-dimensional sphere (the ball's boundary), and $D^{n-1}$ is an $(n-1)$-dimensional ball inside the sphere, allows us to prove that the functors $H$ and $\mathcal{H}$ coincide on any simplex $\Delta^n$, on its boundary $\partial\Delta^n$, and on pairs of form $(\Delta^n, \partial\Delta^n)$, see the solution of Exercise 44. The fact that $H$ and $\mathcal{H}$ coincide on all simplicial complexes can now be established with the help of the following statement, which is called the Five-Lemma.

**Five-Lemma.** *Suppose that the commutative diagram*

$$\begin{array}{ccccccccc}
A & \xrightarrow{i} & B & \xrightarrow{j} & C & \xrightarrow{k} & D & \xrightarrow{l} & E \\
\downarrow{\scriptstyle\alpha} & & \downarrow{\scriptstyle\beta} & & \downarrow{\scriptstyle\gamma} & & \downarrow{\scriptstyle\delta} & & \downarrow{\scriptstyle\varepsilon} \\
A' & \xrightarrow{i'} & B' & \xrightarrow{j'} & C' & \xrightarrow{k'} & D' & \xrightarrow{l'} & E'
\end{array}$$

*of Abelian groups and their homomorphisms is such that its upper and lower rows are exact sequences and the homomorphisms $\alpha$, $\beta$, $\delta$, $\varepsilon$ are isomorphisms. Then $\gamma$ is also an isomorphism.*

**Exercise 46.** Prove this statement.

Applying this lemma to the relevant parts of the long exact sequence of a pair $(K, L)$, where $K$ is a simplicial complex and $L$ is obtained from $K$ by removing the interior of some simplex $\Delta^n$ of maximal dimension, it is easy to show that $H(L) = \mathcal{H}(L) \Longrightarrow H(K) = \mathcal{H}(K)$. It remains to notice that any simplicial complex can be obtained from the point by successive attachment of new simplices.

## 1.12 Digression to the theory of Abelian groups

We start by recalling some necessary material from the theory of finitely generated Abelian groups. These groups are usually defined by indicating their *generators and relations*. Such a definition of an Abelian group has the form $\langle a_1, \dots, a_n \mid R_1, \dots, R_m \rangle$ and consists of a list of generators $a_1, \dots, a_n$ and a list of relations $R_1, \dots, R_m$. Each relation $R_i$ is a formal linear combination of the generators: $R_i = k_{i1}a_1 + \dots + k_{in}a_n$. If the coefficients in all these sums are written as a matrix, then the *relation matrix* thus obtained (which has $m$ rows and $n$ columns) fully determines the particular presentation as well as the group. For instance, the presentation $\langle a, b, c \mid 2a + 3b - c, b + 2c, a + 3b, a + 4b + 2c \rangle$ is determined by its matrix

$$\begin{pmatrix} 2 & 3 & -1 \\ 0 & 1 & 2 \\ 1 & 3 & 0 \\ 1 & 4 & 2 \end{pmatrix}.$$

In order to give a precise description of the group $G$ defined by some presentation $\langle a_1, \dots, a_n \mid R_1, \dots, R_m \rangle$, we proceed as follows. Introduce on the set of all integer linear combinations of the generators $a_j$, $1 \leqslant j \leqslant n$, an equivalence relation stating that two combinations are equivalent if one can be obtained from the other by adding or subtracting the combinations $R_i$, $1 \leqslant i \leqslant m$. The same combination $R_i$ may be added or subtracted several times, or it may not be used at all. Then the elements of $G$ are classes of equivalent combinations. The addition in this group is of course coordinate-wise. Namely, to get the sum of two equivalence classes (*i.e.* of two group elements), we should choose one linear combination from each class, sum up the coefficients at the coinciding generators, and take the equivalence class containing the resulting linear combination.

**Exercise 47.** Prove that the above relation is indeed an equivalence relation, that the coordinate-wise addition is well defined (*i.e.* does not depend on the choice of particular representatives), and that it endows the set of all equivalence classes with the structure of an Abelian group.

The reader has probably noticed already that we have described nothing other than the construction of the quotient group of a certain *free Abelian group* $n\mathbb{Z} = \mathbb{Z} \oplus \dots \oplus \mathbb{Z}$ by its subgroup generated by the elements $R_i$.

**Exercise 48.** What happens to the relation matrix when one of the relations is replaced by its sum with another one? (The group remains unchanged by this operation.)

**Exercise 49.** What happens to the relation matrix if some pair of its generators $a, b$ is replaced by the new pair $a_1 = a + b$ and $b_1 = b$, with appropriate replacements in all the relations? (The group remains unchanged by this operation.)

**Exercise 50.** What happens to the relation matrix if we change the sign of a relation? If we replace a generator $a$ by $a_1 = -a$? If we change the order of generators or that of relations?

The results of these three exercises show that the group remains unchanged under the following elementary transformations of the matrix: adding one row of the matrix to another one, permuting the rows, changing the signs of all the entries in a row, and corresponding transformations of columns.

One can show that any integer matrix $A = (a_{ij})$, $1 \leqslant i \leqslant m$, $1 \leqslant j \leqslant n$, can be brought, by an application of the above transformations, to a diagonal form, *i.e.* the form in which all the non-zero entries are placed on some starting segment of the diagonal $a_{11}, \dots, a_{kk}$ with $k \leqslant \min(m, n)$. Let us observe this in an example.

$$\begin{pmatrix} 2 & 3 & -1 \\ 0 & 1 & 2 \\ 1 & 3 & 0 \\ 1 & 4 & 2 \end{pmatrix} \longrightarrow \begin{pmatrix} 0 & -3 & -1 \\ 0 & 1 & 2 \\ 1 & 3 & 0 \\ 0 & 1 & 2 \end{pmatrix} \longrightarrow \begin{pmatrix} 1 & 3 & 0 \\ 0 & 1 & 2 \\ 0 & -3 & -1 \\ 0 & 1 & 2 \end{pmatrix}$$

$$\longrightarrow \begin{pmatrix} 1 & 3 & 0 \\ 0 & 1 & 2 \\ 0 & 0 & 5 \\ 0 & 0 & 0 \end{pmatrix} \longrightarrow \begin{pmatrix} 1 & 0 & 0 \\ 0 & 1 & 0 \\ 0 & 0 & 5 \\ 0 & 0 & 0 \end{pmatrix}$$

It follows that the group $\langle a, b, c \mid 2a+3b-c, b+2c, a+3b, a+4b+2c\rangle$ is isomorphic to the group $\langle a_1, b_1, c_1 \mid a_1, b_1, 5c_1\rangle$, *i.e.* the group $\mathbb{Z}_1 \oplus \mathbb{Z}_1 \oplus \mathbb{Z}_5 = \mathbb{Z}_5$, since $\mathbb{Z}_1 = 0$.

One can also show that any integer matrix can be brought, by elementary transformations, to the *canonical* diagonal form in which all the diagonal elements are non-negative and each diagonal element $a_{ii}$ is divisible by all the preceding diagonal elements $a_{jj}$, $j < i$. The canonical diagonal form of a matrix does not depend on the particular sequence of transformations (which is why it is called canonical).

To simplify the calculations it is useful to keep in mind the following easy trick. Assume that the matrix arising at some step contains a row composed of all zeros and exactly one unity. Then the group stays the same if we remove from the matrix the "cross" formed by this row and the column containing the unity entry. Similar removal can be done also in the case when there is a column all of whose entries but one are zero and the only non-zero entry is unity.

**Exercise 51.** Present the group $\langle a, b, c \mid a + b + c, a - b + 3c, 2a - 4c\rangle$ as a direct sum of cyclic groups.

**Exercise 52.** Prove that in the case when the relation matrix is a square one, the absolute value of its determinant is equal either to the order of the group or, if the group is infinite, to zero.

## 1.13 Calculation of homology groups

In this section we describe a method for calculating the homology groups of a *free* chain complex $C$,

$$\cdots \longrightarrow C_{n+1} \xrightarrow{\partial_{n+1}} C_n \xrightarrow{\partial_n} C_{n-1} \longrightarrow \cdots,$$

*i.e.* a complex all of whose chain groups are free and finitely generated. Precisely complexes of this type are the chain complexes assigned to simplicial ones. We assume that there is a basis fixed in each chain group and that each boundary homomorphism $\partial_n\colon C_n \to C_{n-1}$ is given by its matrix $A_n$. The matrices $A_n$ are defined according to the standard rule: take the $i$-th base vector of the group $C_n$, apply to it the relevant boundary homomorphism, present the resulting element of the group $C_{n-1}$ as a linear combination of its base elements, and write into the $i$th column the coefficients of this linear combination. Thus, the matrix $A_n$ has $r_{n-1}$ rows and $r_n$ columns, where $r_i$ denotes the free rank of the group $C_i$.

Let us consider what happens to the matrix $A_n$ when the basis of $C_n$ is subjected to one of the following elementary transformations:

1. Adding one base element to another one.
2. Replacing a base element by its inverse.
3. Permuting two base elements.

The answer is simple: the matrix is subjected to the analogous transformations of columns. If an elementary transformation is applied to the basis of the group $C_{n-1}$, then $A_n$ undergoes a certain transformation of rows. Reversing these observations yields an easy way of calculating the $n$th homology group of the complex $C$. Its essence is to perform simultaneous transformations of the matrices $A_{n+1}$ and $A_n$.

*Step* 1. Apply to $A_n$ elementary transformations of rows and columns bringing it to a diagonal form. During this process, whenever we perform an elementary transformation of columns of $A_n$ (which is necessarily induced by a change of basis in $C_n$), we must at the same time perform the corresponding transformation of rows of the matrix $A_{n+1}$ (which is induced by the same change of basis in $C_n$). The correspondence rule is as follows: if the $i$-th column of $A_n$ is added to the $j$-th one, then the $j$-th row of $A_{n+1}$ must be subtracted from the $i$-th row. Permutation of columns of $A_n$ or the change of sign in a column are equivalent to the same respective transformations of rows of $A_{n+1}$.

Denote by $k$ the rank of $A_n$, *i.e.* the length of its maximal non-zero segment of the diagonal. Since the boundary homomorphisms $\partial_{n+1}$, $\partial_n$ of the complex $C$ satisfy the condition $\partial_n\partial_{n+1} = 0$, the product $A_nA_{n+1}$ of their matrices is a zero matrix. It follows that the first $k$ rows of the (new) matrix $A_{n+1}$ consist of zeros only. See Figure 25, where the black cells indicate non-zero entries, white cells indicate zero entries, and grey cells could mean anything. We have written the complex from right to left in order to place the matrices of boundary homomorphisms in the order needed for multiplication according to the usual "row by column" rule.

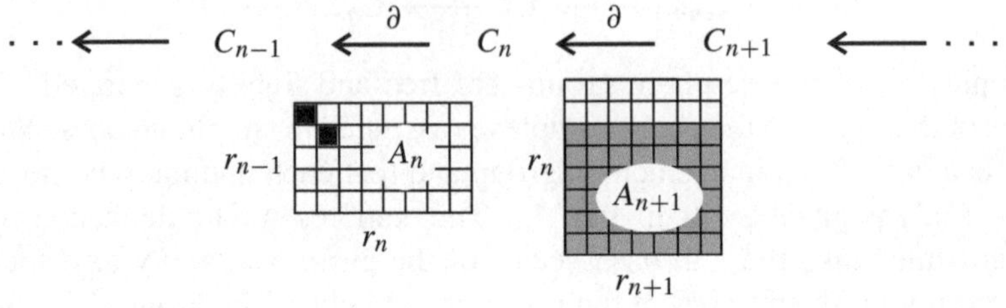


Figure 25. Since $A_n$ is diagonal, the first rows of $A_{n+1}$ consist of zeros.

*Step* 2. We cross out these $k$ rows and bring the matrix $A'_{n+1}$ thus obtained to a diagonal form by elementary transformations of rows and columns. Now the transformations of rows do not need to be accompanied by transformations of columns of $A_n$, because the columns of $A_n$ corresponding to the rows of $A'_n$ consist of zeros only. It is convenient to make all diagonal elements non-negative.

*Step* 3. Now the answer is written as follows: $H_n(C) = \mathbb{Z}_{a_{11}} \oplus \cdots \oplus \mathbb{Z}_{a_{kk}} \oplus s\mathbb{Z}$, where $a_{11}, \ldots, a_{kk}$ are the non-zero diagonal elements of $A'_{n+1}$ and $s$ is the number of zero rows in it. Of course, the summands with $a_{ii} = 1$ can be ignored.

**Exercise 53.** Calculate the homology groups of the chain complex all of whose homology groups are zero except the groups $C_0 = 2\mathbb{Z}$, $C_1 = 4\mathbb{Z}$, $C_2 = 3\mathbb{Z}$, $C_3 = \mathbb{Z}$, and the boundary homomorphisms $\partial_1$, $\partial_2$, $\partial_3$ are defined as follows: $\partial_3$ maps the whole group $C_3$ to zero, $\partial_1$ is given by the $(2 \times 4)$-matrix composed of the rows $(1\ 1\ 1\ 1)$ and $(-1\ -1\ -1\ -1)$, and the matrix of $\partial_2$ is composed of the rows $(1\ 1\ 1)$, $(1\ -1\ -1)$, $(-1\ -1\ 1)$, and $(-1\ 1\ -1)$.

The process of calculating the homology groups described above is useful also from a theoretical point of view. Recall that the elementary chain complex $E(m)$ consists of zero groups and exactly one $\mathbb{Z}$ in dimension $m$ (see Exercise 16). The elementary chain complex $D(m,k)$, $k \neq 0$, has two non-zero groups: in dimension $m$ and in dimension $m+1$. Both groups are isomorphic to $\mathbb{Z}$, and the boundary homomorphism $\partial_{m+1} \colon \mathbb{Z} \to \mathbb{Z}$ is the multiplication by $k$.

**Theorem 21.** *Any chain complex whose chain groups are all free, have finite ranks, and are trivial in negative dimensions, is isomorphic to a direct sum of elementary chain complexes of form $E(m)$ and $D(m,k)$.*

*Proof.* The above method is sufficient for bringing matrices $A_i$ of all the boundary homomorphisms $\partial_i$ to a diagonal form. First we transform $A_1$, then $A_2$, etc. The transformations are performed by changing bases of the chain groups, *i.e.* by choosing different decompositions of these groups into direct sums of $\mathbb{Z}$'s. It remains to notice that all matrices $A_i$ being diagonal means precisely that the corresponding bases determine a decomposition of the complex into a direct sum of complexes of form $E(m)$ and $D(m, k)$. □

To better understand the information carried by the homology groups of a given chain complex $C$, it is useful to present it as the sum $T \oplus H$ where the chain complex $T$ is the sum of all the elementary summands of form $D(m, k)$ with $k = 1$, and $H$ is the sum of the elementary summands of form $E(m)$ and $D(m, k)$ with $k \neq 1$. All the homology groups of $T$ are trivial, while the homology groups of $H$ coincide with those of $C$. The most important point is that the complex $H$ is completely determined by its homology groups. Indeed, to recover $H$ it suffices to present the given homology groups as the sums of non-zero cyclic groups and to replace each cyclic group by the corresponding elementary complex of suitable dimension.

**Exercise 54.** Prove that any sequence $H_0, H_1, \ldots$ of finitely generated Abelian groups can be realized as a sequence of the homology groups of some free chain complex.

## 1.14 Cellular homology

The method of calculating the homology groups described above is quite efficient. At least, it is easily realizable by a computer program. However, calculations by hand are rather difficult, since the number of simplices is usually quite large. For instance, the smallest triangulation of the torus involves 14 triangles. It is however possible to drop the condition that any two triangles either have no common points or intersect along a single vertex or a single edge. Then the torus can be decomposed into two triangles. The above calculation method works for such a *singular triangulation* as well. But then, why not make the next step and represent the torus as a square whose opposing sides are identified, and avoid decomposing the square?

To realize this idea of the most economic decomposition of a polyhedron into simple pieces we recall the notion of the *cell complex*. We prefer to give an inductive definition.

**Definition.**

1. A zero-dimensional cell complex is the union of several points (vertices).

2. A one-dimensional cell complex is obtained from a zero-dimensional one by gluing of several arcs (one-dimensional cells).

3. A two-dimensional complex is obtained from a one-dimensional complex by gluing several two-dimensional cells (discs) along some maps of their boundary curves (see Figure 26).

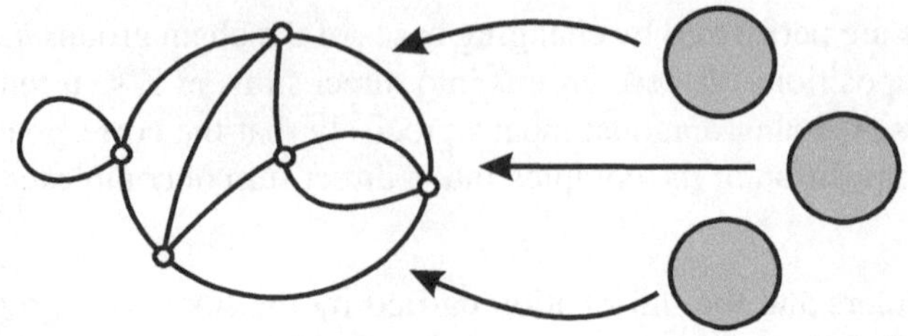

Figure 26. A two-dimensional cell complex is obtained from a one-dimensional complex (a graph) by gluing two-dimensional cells.

4. An $n$-dimensional cell complex is obtained from an $(n-1)$-dimensional one by gluing several $n$-dimensional cells (balls of dimension $n$) along some maps of their boundary spheres.

If $X$ is a cell complex, then the union of all cells of dimension $\leqslant k$ is called its *$k$-dimensional skeleton* and is denoted by $X^{(k)}$. To ensure that a given cell complex is a polyhedron, we always assume that all its skeleta are triangulated and that the gluing maps of all its cells are simplicial.

The construction of the cellular homology theory closely resembles that of the simplicial homology theory. Let $X$ be a cell complex. Orient all its cells and assign to it the chain complex $C(X)$ obtained as follows. For each $n$ the $n$-dimensional chain group $C_n(X)$ is the free Abelian group freely generated (in the natural sense) by all the $n$-dimensional cells. Its elements are formal linear combinations of the form $k_1a_1+\dots+k_ma_m$, where $a_1,\dots,a_m$ are all the $n$-dimensional cells of $X$. To describe the boundary homomorphisms, we need the notion of the *incidence coefficient* of cells.

**Definition.** Let $a$ be an $(n-1)$-dimensional cell of a cell complex $X$, and let $\varphi\colon \partial D^n \to X^{(n-1)}$ be the gluing map of an $n$-dimensional cell $b$. It induces the map $\bar\varphi = p\varphi\colon \partial D^n \to S^{n-1}$ where $p\colon X^{(n-1)} \to S^{n-1}$ is obtained by compressing the $(n-2)$-dimensional skeleton of the $(n-1)$-dimensional skeleton of $X$ and all its cells of dimension $n-1$ except $a$, to a point. Then the *incidence coefficient* $[b:a]$ of the cells $a$ and $b$ is equal to the degree of the map $\bar\varphi$.

The incidence coefficient has a simple geometric meaning. It shows how many times the boundary of the cell $b$ passes along the cell $a$. To calculate it we should choose an $(n-1)$-dimensional simplex in the cell $a$ and count how many $(n-1)$-dimensional simplices contained in the boundary of $D^n$ are mapped to it preserving the orientation and how many are mapped reversing it. The incidence coefficient is equal to the difference of the numbers thus obtained. Notice that it is a natural generalization of the incidence coefficient for simplices $\varepsilon_i$ (see page 10). Therefore it is not surprising that the boundary homomorphisms $\partial_n : C_n \to C_{n-1}$ are given by the similar formulae:

$$\partial_n(b) = \sum_{a_i \in X} [b : a_i] a_i,$$

where the sum is taken over all the $(n-1)$-dimensional cells of $X$.

Furthermore, it is easy to show (see [6]) that for any $n$ we have the equality $\partial_n \partial_{n+1} = 0$, *i.e.* that the boundary homomorphisms $\partial_n$ thus introduced turn the sequence of groups $C_n(X)$ into a chain complex.

**Definition.** The homology groups of the chain complex $C(X)$ are called the *homology groups* of the cell complex $X$ and are denoted by $H_n(X)$.

**Theorem 22.** *For any polyhedron presented as a cell complex, its cellular homology groups coincide with the simplicial ones.*

*Proof.* There are two possible approaches to proving this theorem. The first one has a more conceptual nature but is rather cumbersome. It consists in first stating and proving a theorem on cellular approximation of maps between cell complexes, then describing the induced homomorphisms of the homology groups, and finally verifying that the obtained functor from the category of polyhedra to the category of sequences of Abelian groups satisfies the axioms I–IV.

The second approach is more straightforward. Let us triangulate the given cell complex $X$ and assign to each $n$-dimensional cell the chain composed from all the $n$-dimensional simplices contained in this cell. The coefficient at each simplex is equal to $\pm 1$, depending on whether the orientations of the simplex and of the cell are consistent or not. Then we should verify that the arising maps from the cellular chain groups to the simplicial ones induce isomorphisms of the homology groups in each dimension. □

The method of calculating the homology groups of simplicial complexes described in Section 1.13 works also in the case of cell complexes. For the reader's convenience we describe an easy modification of it, which is very well suited for calculating the first homology group of any cell complex.

**Calculating $H_1$ of a cell complex.** Let $X$ be a cell complex. The group $H_1(X)$ is of course fully determined by its two-dimensional skeleton $X^{(2)}$, which is obtained from the one-dimensional skeleton $\Gamma = X^{(1)}$ by gluing to it several two-dimensional cells.

*Step* 1. Choose a *maximal tree* in the graph $\Gamma$, *i.e.* a subgraph without cycles which contains all the vertices of $\Gamma$. Orient the remaining edges and mark them by letters.

The same thing can be done in a slightly different way. Orient the edges of the graph and cut them one-by-one keeping the graph connected. Each next cut of an edge is denoted by a new letter. The process stops when no further cut is possible.

*Step* 2. Write a matrix whose rows show how many times the boundary curves of the two-dimensional cells pass along the cut edges (taking into account the direction). This matrix is a relation matrix for the first homology group.

*Step* 3. Transform this matrix into a diagonal form and write down the answer as explained in Section 1.12.

**Example.** Suppose that a two-dimensional cell complex $X$ is obtained from the graph shown in Figure 27, left, by gluing five two-dimensional cells whose boundary curves pass along the edges according to the following pattern: $\{1\ 2\ 3\ 4\ 5\}$ (first curve), $\{1\ 1\ -5\ 8\ -2\}$ (second curve), $\{2\ 6\ 7\ 5\}$ (third curve), $\{3\ -6\ 3\ 7\ 8\}$ (fourth curve), $\{4\ -7\ 4\ 8\ 6\}$ (fifth curve). One of the possible results of performing the first step is shown in the middle of the figure. The matrix arising at the second step is shown on the right. Transforming it to the canonic form (and removing superfluous rows and columns), we get the matrix $\left(\begin{smallmatrix}2&0\\0&6\end{smallmatrix}\right)$, which yields $H_1(X) = \mathbb{Z}_2 \oplus \mathbb{Z}_6$.

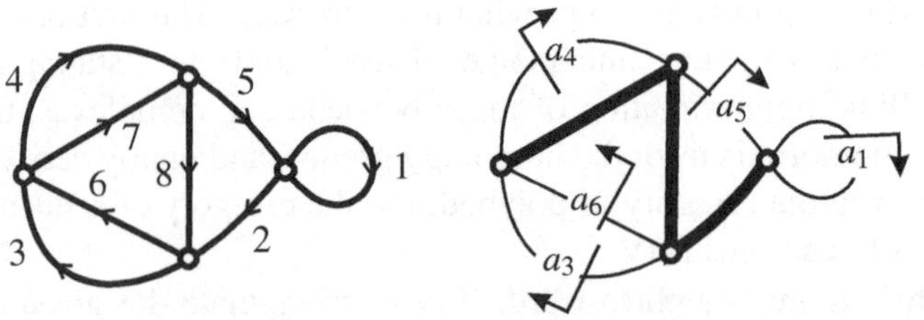


| $a_1$ | $a_3$ | $a_4$ | $a_5$ | $a_6$ |
|---|---|---|---|---|
| 1 | 1 | 1 | 1 | |
| 2 | | | –1 | |
| | | | 1 | 1 |
| | 2 | | | –1 |
| | | 2 | | 1 |

Figure 27. Choosing generators and writing down a relation matrix of the first homology group.

**Exercise 55.** Calculate the first homology group of the Klein bottle.

Along with simplicial and cellular homologies, one can use homologies of other types, for instance, *singular ones.* The difference between the singular homology and the simplicial and cellular ones is in the method of assignment of a chain complex to a given space. The homology groups however are the same. The chain complex in the singular homology theory is generated by *singular simplices, i.e.* mappings of the standard simplex to the space. In non-degenerate cases (when the space consists of an infinite number of points) the set of singular simplices is infinite and often has cardinality continuum. This situation is resolved by requiring that each chain be a linear combination of only a finite number of singular simplices. The singular homology functor satisfies the axioms I–IV, so by Theorem 20 the singular homology groups of polyhedra coincide with the simplicial ones.

The advantage of singular homology groups is that they are well defined for any space. Many theorems that are quite complicated in the simplicial theory become very simple in the singular one. For instance, the theorem on isomorphism of the homology groups of homeomorphic spaces is trivial in the frame of the singular theory. The disadvantages of the singular homology groups are the difficulty of calculating them and the psychological discomfort of operating with infinitely generated groups.

## 1.15 Lefschetz fixed point theorem

Let $f: K \to K$ be a simplicial map from a simplicial complex $K$ to itself. We assume that the fixed point set $L_f$ of $f$ is a subcomplex of $K$. This is easy to achieve, it is sufficient to replace $K$ with its first barycentric subdivision. Then all the invariant (*i.e.* those mapped to itself) simplices of the new complex are fixed point-wise. It turns out that the induced homomorphisms $f_*: H_n(K) \to H_n(K)$ of the homology groups carry essential information about the structure of the set $L_f$.

Each endomorphism (a homomorphism to itself) $\varphi: A \to A$ of a finitely generated Abelian group induces an endomorphism $\varphi': \mathrm{Free}(A) \to \mathrm{Free}(A)$ of the free Abelian group $\mathrm{Free}(A) = A/\mathrm{Tor}(A)$, where $\mathrm{Tor}(A)$ is the *torsion subgroup* of $A$, which consists of all the finite-order elements. The group $\mathrm{Free}(A)$ can be interpreted as the free part of $A$. Since $\mathrm{Free}(A)$ is free, it has a free basis $a_1, \dots, a_n$. Therefore the endomorphism $\varphi'$ can be given by an integer square matrix $M = (m_{ij})$ of order $n$. The matrix $M$ is composed in the standard way: its columns consist of the coordinates of the images of the base elements.

**Definition.** The *trace* $\mathrm{Tr}(\varphi)$ of the endomorphism $\varphi$ is the trace of the matrix $M$, *i.e.* the sum of all its diagonal elements.

It should be stressed that we completely ignore the periodic elements of $A$ when calculating the trace of an endomorphism. The trace depends only on the behaviour of the free part.

**Exercise 56.** Prove that the trace of $\varphi$ is well defined (*i.e.* does not depend on the choice of a basis of $\mathrm{Free}(A)$).

**Exercise 57.** Prove that the trace is additive in the following sense. Let $\varphi: A_1 \oplus A_2 \to A_1 \oplus A_2$ be an arbitrary endomorphism of the direct sum of two finitely generated Abelian groups. Then $\mathrm{Tr}(\varphi) = \mathrm{Tr}(\varphi_1) + \mathrm{Tr}(\varphi_2)$, where for $i = 1, 2$ the endomorphism $\varphi_i: A_i \to A_i$ is the composition of the embedding $A_i \to A_1 \oplus A_2$, the embedding $\varphi$, and the projection $A_1 \oplus A_2 \to A_i$.

Let $C$ be a *finitely generated* chain complex, *i.e.* a chain complex where all chain groups $C_n$ are finitely generated and the number of non-zero chain groups is finite. Consider an arbitrary chain endomorphism $\varphi: C \to C$, *i.e.* a chain map to itself. It consists of a family of endomorphisms $\varphi_n: C_n \to C_n$ of the chain groups. Of course, these endomorphisms must commute with the boundary homomorphisms.

**Definition.** The *Lefschetz number* $\lambda(\varphi)$ of the endomorphism $\varphi$ is defined by the formula $\lambda(\varphi) = \sum_{-\infty}^{\infty}(-1)^n \mathrm{Tr}(\varphi_n)$.

The alternating sum over the dimensions appearing in this definition closely resembles the *Euler characteristic* $\chi(K) = \sum_{-\infty}^{\infty}(-1)^n s_n(K)$ of a finite simplicial complex $K$, where $s_n(K)$ denotes the number of $n$-dimensional simplices of $K$. As the statements of the following two exercises show, this resemblance is not by accident.

**Exercise 58.** Prove that $\chi(K)$ is equal to the Lefschetz number of the identity map of the chain complex $C(K)$ onto itself (recall that the chain groups of $C(K)$ are generated by the simplices of $K$).

**Exercise 59.** Let $f\colon K \to K$ be a simplicial map of a simplicial complex $K$ such that on each invariant simplex it is the identity. Prove that the Lefschetz number of the induced chain map $f_*\colon C(K) \to C(K)$ coincides with the Euler characteristic of the fixed point set of $f$.

Recall that each chain endomorphism $\varphi\colon C \to C$ induces homomorphisms $(\varphi_*)_n\colon H_n(C) \to H_n(C)$ of the homology groups of $C$.

**Definition.** The *homological Lefschetz number* $\lambda(\varphi_*)$ of the endomorphism $\varphi$ is defined by the formula $\lambda(\varphi_*) = \sum_{-\infty}^{\infty}(-1)^n \operatorname{Tr}((\varphi_*)_n)$.

Despite the superficial resemblance of the formulae, there is a significant difference between the definition of the Lefschetz number and that of the homological Lefschetz number: in the latter case, we pass to the level of the homology. Yet, these numbers always coincide.

**Theorem 23.** *For any chain endomorphism $\varphi\colon C \to C$ the Lefschetz number $\lambda(\varphi)$ coincides with the homological Lefschetz number $\lambda(\varphi_*)$.*

*Proof.* Assume that the complex $C$ is free. Otherwise we could quotient it by the torsion part and pass to a free complex. Since the torsion does not influence the Lefschetz number, both numbers are preserved by this operation.

Theorem 21 implies that the complex $C$ can be presented in the form $C = \oplus_j C^{(j)}$, where each $C^{(j)}$ is an elementary complex of type $E(m)$ or of type $D(m,k)$. Since the trace is additive (see Exercise 57), we have that $\lambda(\varphi) = \sum_j \lambda(\varphi_j)$, where each endomorphism $\varphi_j\colon C^{(j)} \to C^{(j)}$ is the composition of the embedding $C^{(j)} \to C$, the given endomorphism $\varphi\colon C \to C$, and the projection $C \to C^{(j)}$. The concluding step of the proof consists in the direct verification of the conclusion of the theorem for any endomorphism of an elementary chain complex of type $E(m)$ or of type $D(m,k)$. □

**Exercise 60.** Prove that the Euler characteristic $\chi(K) = \sum_{-\infty}^{\infty}(-1)^n s_n(K)$ of a finite simplicial complex $K$ is equal to the alternating sum $\sum_{-\infty}^{\infty}(-1)^n r(H_n(K))$, where $r(H_n(K))$ denotes the rank of $H_n(K)$.

Let us switch to geometry now. Let $f\colon P \to P$ be an arbitrary map from a polyhedron $P$ to itself. It induces endomorphisms $(f_*)_n\colon H_n(P) \to H_n(P)$ of its homology groups.

**Definition.** The *Lefschetz number $\lambda(f)$ of $f$* is defined by the formula

$$\lambda(f) = \sum_{-\infty}^{\infty} (-1)^n \operatorname{Tr}((f_*)_n).$$

Of course, if $f$ is simplicial with respect to some triangulation $K$ of the polyhedron $P$, then the number $\lambda(f)$ coincides with the homological Lefschetz number of the induced map $f_*\colon C(K) \to C(K)$.

**Theorem 24.** *Let $L_f$ be the fixed point set of a simplicial map $f\colon K \to K$ from a finite simplicial complex $K$ to itself. Then $\chi(L_f) = \lambda(f)$.*

*Proof.* The proof follows from the result of Exercise 59 and from Theorem 23. □

**Example.** Let $r, r', r''$ be three symmetries of the standard two-sphere $S^2$, with respect to the center, with respect to a diameter axis, and with respect to a diameter plane, see Figure 28. Their Lefschetz numbers are equal to 0, 2, and 0, respectively (since $r$ and $r''$ reverse the orientation of the sphere and therefore induce the multiplication by $-1$ in the group $H_2(S^2) = \mathbb{Z}$). The Euler characteristics of the fixed point sets of these symmetries (the empty set, two points, a circle) are also equal to 0, 2, and 0, in complete agreement with the statement of the theorem.

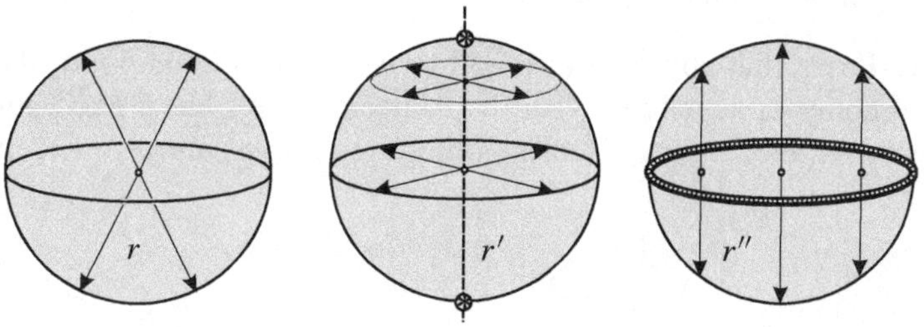


Figure 28. Three symmetries of the standard sphere.

It would be appropriate to emphasize here a small subtlety regarding the possible applications of Theorem 24. The point is that in applications the given map $f$ is rarely simplicial. Of course, it can be approximated by a simplicial one, but this would change the map, which may significantly alter the structure of the fixed point set. However, the following *Lefschetz theorem* works very well even in case of arbitrary maps.

**Theorem 25.** *Let $f\colon P \to P$ be a map from a compact polyhedron $P$ to itself. If $\lambda(f) \neq 0$, then $f$ has at least one fixed point.*

*Proof.* Let $\rho$ be some metric on $P$. For instance, we could take the usual distance in the Euclidean space $\mathbb{R}^N$ containing $P$. Arguing by contradiction, assume that there are no fixed points. Then there exists a number $\varepsilon > 0$ such that for any $x \in P$ there is the inequality $\rho(x, f(x)) > \varepsilon$. This follows from compactness of $P$.

Choose now a triangulation $K$ of $P$ such that the diameter of each simplex is less than $\varepsilon/2$. By the simplicial approximation theorem (Theorem 6) there exist a subdivision $K_1$ of $K$ and a simplicial map $g\colon K_1 \to K$ such that points $f(x)$ and $g(x)$ always belong to the same simplex and therefore the distance between them is always less than $\varepsilon/2$.

Recall that the homology groups of $K_1$ and $K$ are naturally isomorphic. The natural isomorphism between them is induced by the chain map $\alpha\colon C(K) \to C(K_1)$ defined by assigning to each oriented simplex $\sigma^n$ in $K$ the $n$-dimensional chain of $K_1$ composed of all the coherently oriented $n$-dimensional simplices of $K_1$ that constitute $\sigma$. See the discussion of a similar question on page 12.

Now consider the composition $\alpha g_*\colon C(K_1) \to C(K_1)$, where the chain map $g_*\colon C(K_1) \to C(K)$ is induced by $g$. By Theorem 23 the Lefschetz number of this composition map is equal to its homological Lefschetz number, which in turn is equal to $\lambda(f)$. Since $\lambda(f) \neq 0$ by assumption, at least for one dimension the matrix of the respective endomorphism $\alpha g_*$ contains a non-zero diagonal element. It follows that there is at least one simplex $\delta$ of the complex $K_1$ that is contained in the simplex $g(\delta)$. Choosing an arbitrary point $x \in \delta$, we get that $\rho(x, g(x)) < \varepsilon/2$. On the other hand, since the map $g$ is an $\varepsilon/2$-approximation of $f$, the inequality $\rho(f(x), g(x)) < \varepsilon/2$ holds for all points. Together these two inequalities yield the inequality $\rho(x, f(x)) < \varepsilon$, which contradicts the choice of the number $\varepsilon$. □

**Exercise 61.** Prove that any map of a ball to itself has a fixed point. Is it true that any map of a compact acyclic polyhedron to itself always has a fixed point as well? (A polyhedron is called *acyclic* if all its homology groups are zero except the group $H_0 = \mathbb{Z}$.)

## 1.16 Homology with coefficients

Let $G$ be an arbitrary Abelian group. The homology theory with coefficients in $G$ differs from the integer theory only in the dimension axiom whose statement is obtained by replacing the group $\mathbb{Z}$ by $G$:

$$H_n(*) = \begin{cases} 0, & n \neq 0, \\ G, & n = 0. \end{cases}$$

The construction of the homology theory with coefficients in $G$ does not essentially differ from the construction of the homology theory with integer coefficients (see Section 1.4); therefore we restrict ourselves to a brief recap. Let $K$ be an oriented simplicial complex. Its $n$-dimensional chain group $C_n(K; G)$ consists of linear combinations of the form $g_1\sigma_1 + g_2\sigma_2 + \dots + g_k\sigma_k$, where the $\sigma_i$ are $n$-dimensional simplices, and this time the coefficients $g_i$ are not integers but rather elements of $G$. As before, the addition is coordinate-wise, and with respect to it the set $C_n(K;G)$ is a group.

The value of the boundary homomorphisms $\partial_n : C_n(K;G) \to C_{n-1}(K;G)$ on a simplex multiplied by a coefficient is given by the formula

$$\partial_n(g\sigma) = \sum_{\delta_i \in K} \varepsilon_i g \delta_i .$$

The homomorphisms are then extended by additivity. As before, the sum is taken over all the simplices $\delta_i$ of dimension $n-1$, and $\varepsilon_i$ are the incidence coefficients. The arising sequence of Abelian groups and homomorphisms

$$\cdots \longrightarrow C_{n+1}(K;G) \xrightarrow{\partial_{n+1}} C_n(K;G) \xrightarrow{\partial_n} C_{n-1}(K;G) \longrightarrow \cdots$$

satisfies the relations $\partial_n\partial_{n+1} = 0$ and therefore is a chain complex. We denote it by $C(K;G)$. Its homology groups are denoted by $H_n(K;G)$ and are called the *homology groups of $K$ with coefficients in $G$*.

The further construction of the homology theory with coefficients in $G$ does not differ from the case of integer coefficients. In particular, the definitions of induced homomorphisms and of relative homology groups remain the same. The long exact sequence of a pair of spaces and the Mayer–Vietoris exact sequence, as well as all axioms (except the dimension axiom), are also preserved.

Homology theories with coefficients in various Abelian groups can frequently be more convenient than the integer homology. For instance, in the group $\mathbb{Z}_2$ the elements 1 and $-1$ coincide, so we do not need to keep track of orientations of simplices and can consider unoriented complexes. In this case many definitions become noticeably simpler. For instance, a cycle with coefficients in $\mathbb{Z}_2$ is a collection of $n$-dimensional simplices such that each simplex of dimension $n-1$ is adjacent to an even number of simplices of the collection. If the coefficient group is a field $\mathbb{F}$ of characteristic 0 (for example, the rational field or that of real numbers) then there is no torsion and any homology group has the form $\mathbb{F} \oplus \dots \oplus \mathbb{F}$, *i.e.* is completely determined by its rank.

However, the following natural question arises: what is the relation between the homology groups with coefficients in an arbitrary Abelian group $G$ and those with coefficients in $\mathbb{Z}$? The answer is given by the so-called *Universal Coefficient Theorem*.

**Theorem 26.** *For any simplicial complex $K$ and for any $n$ there is the equality $H_n(K;G) = H_n(K) \otimes G \oplus H_{n-1}(K) * G$.*

The sign $\otimes$ denotes the usual *tensor product* of two groups. It can be defined as follows. Let $A$, $B$ be two Abelian groups. Then the group $A \otimes B$ is defined as the Abelian group generated by all pairs of form $(a, b)$ with $a \in A$, $b \in B$, considered modulo the bilinearity relations $(a_1+a_2, b) = (a_1, b)+(a_2, b)$ and $(a, b_1+b_2) = (a, b_1)+(a, b_2)$. The tensor product possesses the following properties:

1. $A \otimes B = B \otimes A$;
2. $(A_1 \oplus A_2) \otimes B = A_1 \otimes B \oplus A_2 \otimes B$;
3. $A \otimes \mathbb{Z}_m = A/mA$, where $A/mA$ denotes the quotient group of $A$ by the subgroup consisting of the elements of form $ma$, $a \in A$.

In all three cases the equality sign means the existence of a natural isomorphism.

The last property allows us to calculate easily the tensor product of two cyclic groups: $\mathbb{Z}_k \otimes \mathbb{Z}_m = \mathbb{Z}_{(k,m)}$, where $(k, m)$ is the greatest common divisor of $k$ and $m$. In particular, $\mathbb{Z} \otimes \mathbb{Z}_m = \mathbb{Z}_m$. Together with the distributivity (property 2) this suffices to calculate the tensor product of any two finitely generated Abelian groups.

The appearance of the tensor product in the universal coefficient formula is quite natural. Indeed, considering chain groups with coefficients in $G$ is equivalent to taking the tensor product of the integer chain groups with $G$. In other words, the groups $C_n(K) \otimes G$ and $C_n(K; G)$ are naturally isomorphic. Therefore it is not at all surprising that the groups $H_n(K) \otimes G$ contribute to the groups $H_n(K; G)$. Possibly unexpected "additions" $H_{n-1}(K) * G$ appear due to the fact that $G$ may contain periodic elements.

*The torsion product* $A * B$ of Abelian groups is defined in a slightly more complicated way. Let us write an exact sequence $0 \to F_1 \xrightarrow{i} F_2 \to A \to 0$, where the Abelian groups $F_1$, $F_2$ are free. Such a sequence (it is called a *short free resolution of* $A$) always exists. If we take its tensor product with $B$, then the homomorphism $i$ stops being injective and the kernel appears. This kernel is called the torsion product of the groups $A$ and $B$. Thus, the torsion product is included in the exact sequence $0 \to A * B \to F_1 \otimes B \xrightarrow{i \otimes \mathrm{Id}} F_0 \otimes B \to A \otimes B \to 0$. Some of the properties of the torsion product are similar to those of the tensor product:

1. $A * B = B * A$;
2. $(A_1 \oplus A_2) * B = A_1 * B \oplus A_2 * B$;
3. $A * \mathbb{Z}_m$ is the subgroup consisting of those elements of $A$ whose multiple by $m$ is zero.

However, a new property appears as well: the torsion product of two groups $A$, $B$ depends only on their torsion subgroups $\mathrm{Tor}(A)$, $\mathrm{Tor}(B)$. More precisely, the following relation always holds: $A * B = \mathrm{Tor}(A) * \mathrm{Tor}(B)$. Therefore the groups $\mathbb{Z} * \mathbb{Z}_m$ and $\mathbb{Z}_k * \mathbb{Z}$ are always zero. If some cyclic groups $\mathbb{Z}_k$, $\mathbb{Z}_m$ are finite (*i.e.* if $k, m \neq 0$) then, as in the case of the tensor product, $\mathbb{Z}_k * \mathbb{Z}_m = \mathbb{Z}_{(k,m)}$.

**Exercise 62.** Determine the group $(\mathbb{Z}_2 \oplus \mathbb{Z}_6) * (\mathbb{Z} \otimes \mathbb{Z}_9)$.

*Proof of Theorem* 26. Recall that the complex $C(K)$ can be presented as a direct sum of elementary complexes of type $E(m)$ or $D(m, k)$ (Theorem 21). Therefore the chain complex $C(K; G) = C(K) \otimes G$ can be presented as a sum of complexes of type

$E(m) \otimes G$ and $D(m, k) \otimes G$. When we sum up chain complexes, we should sum up their homology groups as well. Therefore to prove the theorem it suffices to verify that the homology groups of the complexes $E(m) \otimes G$ and $D(m, k) \otimes G$ are expressed through the homology groups of $E(m)$ and $D(m, k)$ precisely as indicated by the universal coefficient formula. □

**Exercise 63.** Find the homology groups with coefficients in $\mathbb{Z}_2$ for the projective space $\mathbb{R}P^3$.

**Exercise 64.** Find the first homology groups of the Klein bottle with coefficients in $\mathbb{Z}$, $\mathbb{Z}_2$, $\mathbb{Z}_3$, and $\mathbb{Q}$, where $\mathbb{Q}$ is the group of rational numbers.

## 1.17 Elements of cohomology theory

In contrast with homology theory, cohomology theory is a contravariant functor. In some sense, these two theories are adjoint (or dual) to each other.

Let $C$ be an arbitrary chain complex

$$\cdots \longrightarrow C_{n+1} \xrightarrow{\partial_{n+1}} C_n \xrightarrow{\partial_n} C_{n-1} \longrightarrow \cdots.$$

For each $n$ we define the *cochain group* $C^n$ as the group of all homomorphisms of the chain group $C_n$ to $\mathbb{Z}$. Thus, each $n$-dimensional cochain is an integer-valued functional on the $n$-dimensional chain group. If $x \in C^n$ and $y \in C_n$ are a cochain and a chain of the same dimension, then we denote by $(x, y)$ the number equal to the value $x(y)$ of the cochain $x$ on the chain $y$. This allows us to define *coboundary homomorphisms* $\delta_n\colon C^{n-1} \to C^n$ by the rule $(\delta_n x, y) = (x, \partial_n y)$, where $x \in C^{n-1}$ and $y \in C_n$.

It should be noticed here that in the case of a free complex $C$, each of its chain groups $C_n$ is isomorphic to the cochain group $C^n$, since both chains and cochains are determined by assigning some integers to the free generators of the chain groups. In the first case these numbers play the rôle of coefficients, in the second, that of the values of the functional. However, there is no canonic isomorphism.

The equality $(\delta_n x, y) = (x, \partial_n y)$ means that the homomorphism $\delta_n$ is *dual* to the homomorphism $\partial_n$. It follows that, since $\partial_n \partial_{n+1} = 0$, we also have $\delta_{n+1}\delta_n = 0$. Therefore the *group of coboundaries* $B^n = \operatorname{Im} \delta_n$ is always contained in the *group of cocycles* $A^n = \operatorname{Ker} \delta_{n+1}$. We see that the cochain complex $C^*$

$$\cdots \longrightarrow C^{n-1} \xrightarrow{\delta_n} C^n \xrightarrow{\delta_{n+1}} C^{n+1} \longrightarrow \cdots$$

is in fact a chain complex, the difference being only in the numbering of groups and homomorphisms. The star in the notation of the cochain complex is added in order to distinguish it from a chain complex and to recall that cochain groups are dual to the relevant chain groups, *i.e.* the former consist of additive functionals on the latter. In this sense it would be natural to denote coboundary homomorphisms by $\partial_n^*$, but we stick to the traditional notation $\delta_n$.

**Definition.** The quotient group $A^n/B^n = \operatorname{Ker} \delta_{n+1}/\operatorname{Im} \delta_n$ is called the *n-dimensional cohomology group* of the complex $C$ and is denoted by $H^n(C)$.

Calculation of the cohomology groups of free finitely generated chain complexes is carried out according to the same scheme as that of the homology groups. The only difference is that the groups of a cochain complex are numbered in the opposite order, *i.e.* coboundary homomorphisms increase the dimension. Matrices of coboundary homomorphisms can be either calculated directly or written down using the property that, if we choose the dual basis, then the matrices of coboundary homomorphisms are obtained by transposing those of the boundary homomorphisms of the chain complex. Then they should be brought to the diagonal form, as described in Section 1.13.

**Definition.** Let $K$ be a simplicial complex. Then the cohomology groups of the corresponding chain complex $C(K)$ are called the *cohomology groups of* $K$ and are denoted by $H^n(K)$.

As we remarked before, the chain groups and the cochain groups of a simplicial complex, viewed as abstract groups, are always isomorphic. The essential distinction between a boundary homomorphism and a coboundary one is that, under the boundary homomorphism $\partial_n$, the coefficients on the $n$-dimensional simplices are carried down (with appropriate signs) to their boundaries, which are then summed up, while under the coboundary homomorphism $\delta_n$ the values on the $(n-1)$-dimensional simplices are summed up (with signs) to form the values assigned to the $n$-simplices bounded by them. See Figure 29

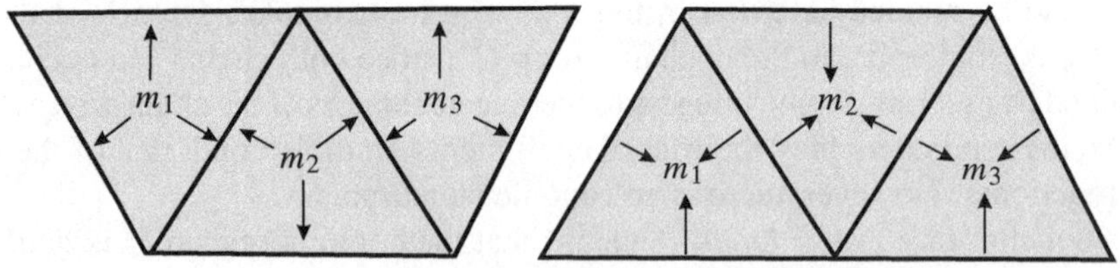


Figure 29. The boundary homomorphisms and the coboundary ones act in different directions.

Each chain map $\varphi\colon C \to C'$ induces homomorphisms $\varphi^*\colon H^n(C') \to H^n(C)$ between the cohomology groups. Therefore each simplicial map $f\colon K \to K'$ between simplicial complexes also induces homomorphisms $f^*\colon H^n(K') \to H^n(K)$ of their cohomology groups. These homomorphisms act in the opposite direction, hence the cohomology theory is a contravariant functor. Just as in the case of homology, one can prove that this functor satisfies the suitably modified homotopy axiom, long exact sequence of a pair axiom, excision axiom, and dimension axiom. In particular, the long exact cohomology sequence of a pair $(K, L)$ of simplicial complexes has the form

$$\cdots \longrightarrow H^{n-1}(L) \xrightarrow{\delta} H^n(K, L) \xrightarrow{p^*} H^n(K) \xrightarrow{i^*} H^n(L) \longrightarrow \cdots.$$

Obviously, the homology groups and the cohomology groups of a simplicial complex are closely related. In fact, knowing groups of one type we can find groups of the other type. This is true for any finitely generated free chain complex. Let $A$ be a finitely generated Abelian group. Recall that $\operatorname{Free}(A)$ and $\operatorname{Tor}(A)$ denote the free part $A/\operatorname{Tor}(A)$ and the torsion subgroup of $A$, respectively. The group $A$ is always isomorphic to the group $\operatorname{Free}(A) \oplus \operatorname{Tor}(A)$. However, there are many different isomorphisms, all of them equivalent, meaning that there are no reasons for selecting one preferred (*canonic*) isomorphism.

**Theorem 27.** *The homology groups and the cohomology groups of any finitely generated free chain complex $C$ are related by the equalities* $\operatorname{Free}(H^n(C)) = \operatorname{Free}(H_n(C))$ *and* $\operatorname{Tor}(H^n(C)) = \operatorname{Tor}(H_{n-1}(C))$. *In other words, the cohomology group $H^n(C)$ is isomorphic to the direct sum* $\operatorname{Free}(H_n(C)) \oplus \operatorname{Tor}(H_{n-1}(C))$.

*Proof.* A direct verification shows that the theorem is true for all elementary complexes $E(m)$ and $D(m, k)$. The validity of the general case of the theorem follows from the additivity of the homology groups and the cohomology groups with respect to taking direct sums of chain complexes, and from Theorem 21 stating that each free chain complex can be presented as a direct sum of elementary complexes. □

**Exercise 65.** Calculate the cohomology groups of the Klein bottle in two ways, straight from the definition and using Theorem 27.

It is interesting to note that in serious applications the cohomology groups appear even more often than the homology ones. What is the advantage of cohomology groups? One explanation is that the cohomology groups of a smooth orientable manifold $M$ with coefficients in $\mathbb{R}$ reflect the degree of inexactness of the de Rham complex of differential forms on $M$. Another explanation refers to algebraic considerations: the direct sum $\sum_{n=0}^{\infty} H^n(K)$ of all the cohomology groups of an arbitrary complex $K$ possesses a natural ring structure. To describe it we recall the definition of the tensor product of chain complexes.

**Definition.** Let $X$ and $Y$ be chain complexes. Their *tensor product* is the chain complex $C = X \otimes Y$ whose chain groups are defined by the equality $C_n = \sum_{i+j=n} X_i \otimes Y_j$ and whose boundary homomorphisms are defined by assigning certain values to the elements $a \otimes b \in X_i \otimes Y_j$, $i + j = n$, according to the formula $\partial_n(a \otimes b) = \partial_i a \otimes b + (-1)^i a \otimes \partial_j b$, which can also be written as $\partial(a \otimes b) = \partial a \otimes b + (-1)^{\dim a} a \otimes \partial b$.

**Exercise 66.** Check that the boundary homomorphisms thus defined satisfy the requirement $\partial_n \partial_{n+1} = 0$.

The tensor product of chain complexes has a transparent geometric meaning: it corresponds to the direct product of simplicial complexes. Let $K$, $L$ be arbitrary simplicial complexes. We decompose the direct product $|K| \times |L|$ into cells of the form $\sigma_1 \times \sigma_2$, where $\sigma_1 \in K, \sigma_2 \in L$ are simplices. The cell complex thus obtained is denoted as $K \times L$. Let $C(K \times L)$ be the corresponding chain complex.

**Theorem 28.** *For any simplicial complexes $K, L$ the chain complexes $C(K) \otimes C(L)$ and $C(K \times L)$ are isomorphic. There is a natural isomorphism $\varphi\colon C(K) \otimes C(L) \to C(K \times L)$, which can be given by assigning $\sigma_1 \otimes \sigma_2 \to \sigma_1 \times \sigma_2$, where $\sigma_1 \in K, \sigma_2 \in L$ are simplices and $\sigma_1 \otimes \sigma_2, \sigma_1 \times \sigma_2$ are viewed as elements of chain groups.*

*Proof.* It is easy to check that the homomorphisms of the chain groups defined by the above assignments commute with the boundary homomorphisms and therefore define a chain map $\varphi$ from $C(K) \otimes C(L)$ to $C(K \times L)$. This verification also exhibits the geometric meaning of the factor $(-1)^{\dim a}$ in the definition of the boundary of $a \otimes b$. Without this factor we would not get the correct boundary, see Figure 30. It is obvious that $\varphi$ is surjective, since its image contains all the generators of $C_n(K \times L)$. The injectivity is easily proven by calculating the ranks of $n$-dimensional chain groups of the complexes $C(K) \otimes C(L)$ and $C(K \times L)$: both ranks are equal to the number of $n$-dimensional cells in $K \times L$. □

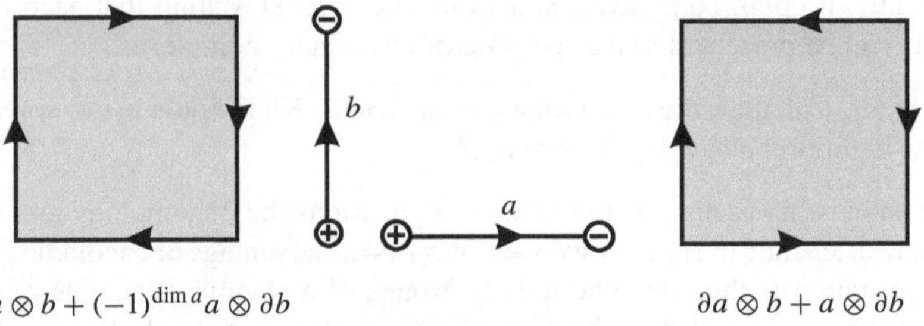


Figure 30. Without the factor $(-1)^{\dim a}$ the orientations of segments of the boundary of the square are wrong.

The tensor product of chain complexes, just as the tensor product of Abelian groups, is symmetric and distributive with respect to taking direct sums, see page 51.

**Exercise 67.** Prove that for any two chain complexes $X$, $Y$ the complex $(X \otimes Y)^*$ is naturally isomorphic to the complex $X^* \otimes Y^*$.

**Theorem 29.** *For any two finitely generated free chain complexes $X$, $Y$, assigning to each pair of cycles $a \in X_k$, $b \in Y_m$ the chain $a \otimes b$ of dimension $k + m$ of the complex $X \otimes Y$ defines embeddings $i\colon \sum_{k+m=n} H_k(X) \otimes H_m(Y) \to H_n(X \otimes Y)$.*

*Proof.* A direct verification shows that the theorem is true for the elementary complexes $E(m)$ and $D(m,k)$. The validity of the theorem in the general case follows from Theorem 21 stating that any free chain complex can be presented as a direct sum of elementary ones. □

In fact, there is a more precise result, which goes under the name of the "Künneth formula" and is proved in the same manner: for any two finitely generated free chain complexes $X$, $Y$ the sequence

$$0 \longrightarrow \sum_{k+m=n} H_k(X) \otimes H_m(Y) \xrightarrow{\ i\ } H_n(X \otimes Y) \xrightarrow{\ p\ } \sum_{k+m=n-1} H_k(X) * H_m(Y) \longrightarrow 0$$

is exact. The homomorphisms $i$ and $p$ in this sequence are functorial, *i.e.* they commute with the homomorphisms of the homology groups induced by the chain maps of the complexes. Moreover, the sequence splits, *i.e.* there exists an opposite homomorphism $s\colon \sum_{k+m=n-1} H_k(X) * H_m(Y) \to H_n(X \otimes Y)$ such that $ps = 1$. It follows that the group $H_n(X \otimes Y)$ is isomorphic to the direct sum of the groups $\sum_{k+m=n} H_k(X) \otimes H_m(Y)$ and $\sum_{k+m=n-1} H_k(X) * H_m(Y)$, but such a decomposition is not functorial.

The following theorem is an analogue of Theorem 29 for cohomology.

**Theorem 30.** *For any two finitely generated chain complexes $X$, $Y$, assigning to each pair of cocycles $a \in X^k$, $b \in Y^m$ the cochain $a \otimes b \in X \otimes Y$ determines an embedding $i\colon \sum_{k+m=n} H^k(X) \otimes H^m(Y) \to H^n(X \otimes Y)$.*

The proof remains the same, since cochain complexes are essentially chain complexes. The difference consists only in the numbering of the groups, which does not influence the validity of the theorem. Certainly, just as for homology, the embedding $i$ is included into the functorial split exact sequence

$$0 \longrightarrow \sum_{k+m=n} H^k(X) \otimes H^m(Y) \xrightarrow{\ i\ } H^n(X \otimes Y) \xrightarrow{\ p\ } \sum_{k+m=n-1} H^k(X) * H^m(Y) \longrightarrow 0,$$

therefore we have

$$H^n(X \otimes Y) = \sum_{k+m=n} H^k(X) \otimes H^m(Y) \oplus \sum_{k+m=n-1} H^k(X) * H^m(Y).$$

Now we are ready to define the cohomology product. Let $K$ be a simplicial complex and let $\Delta\colon |K| \to |K| \times |K|$ be the *diagonal map* given by the formula $\Delta(x) = (x, x)$. It induces homomorphisms $\Delta^*\colon H^n(K \times K) \to H^n(K)$.

**Definition.** Let $h_1 \in H^k(K)$, $h_2 \in H^m(K)$ be two elements of cohomology groups of a simplicial complex $K$. Then their $\cup$-*product* (the *cup product*) $h_1 \cup h_2$ is defined by the formula $h_1 \cup h_2 = \Delta^* i(h_1 \otimes h_2)$.

In other words, in order to find the product of two elements of cohomology groups we should consider their tensor product as an element of the relevant cohomology group of the direct product $K \times K$ and take its image under the homomorphism $\Delta^*$ induced by the diagonal map. However, a direct calculation of the product in cohomology is not an easy task, even if there are no conceptual difficulties. The problem is that a straightforward description of $\Delta^*$ requires finding a simplicial (or cellular) approximation of the diagonal map $\Delta\colon K \to K \times K$. Although there are explicit formulas for such an approximation, they are quite cumbersome.

## 1.18 The Poincaré duality

Let $M$ be a closed orientable manifold of dimension $n$. For simplicity we assume that $M$ is a combinatorial manifold. This means that $M$ can be triangulated in such a way that the closed star of any vertex is simplicially isomorphic to a subdivision of the standard $n$-dimensional simplex. An equivalent definition: $M$ is a *homogenous polyhedron* (this expression is due to S. P. Novikov). This means that any point $x \in M$ can be taken to any other point $y \in M$ by a homeomorphism $M \to M$ that is simplicial with respect to some triangulations.

We choose a triangulation $K$ of $M$ and we consider the *dual decomposition* into cells. Namely, to each $m$-dimensional simplex $\sigma$ of $K$ we assign the dual cell $B_\sigma$ of dimension $n-m$. It is defined as $B_\sigma = \bigcap_{i=0}^m \mathrm{St}(v_i, K')$, where $K'$ is the first barycentric subdivision of $K$ and the intersection is taken over all the vertices $v_i$ of $\sigma$. In particular, to each vertex $v$ of $K$ we assign the $n$-dimensional ball $B_v = \mathrm{St}(v, K')$, to each edge $\sigma$ with endpoints $v$, $w$ the $(n-1)$-dimensional ball $B_\sigma = B_v \cap B_w$, and so forth. See Figure 31.

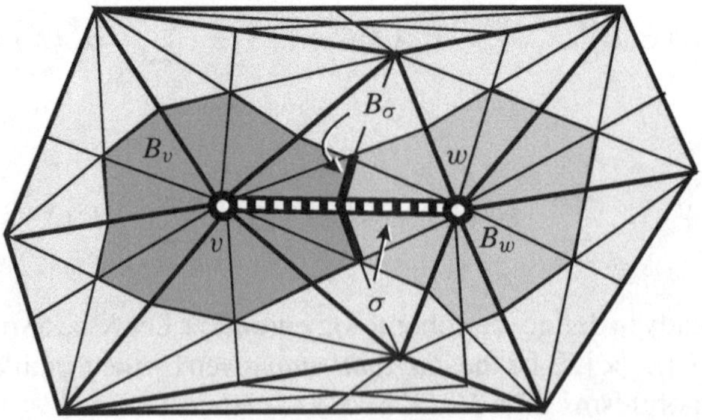


Figure 31. The dual cell decomposition of a manifold.

**Theorem 31.** *For any closed orientable combinatorial manifold $M$ of dimension $n$ and for any $m$, the groups $H_m(M)$ and $H^{n-m}(M)$ are isomorphic.*

*Proof.* Let $K$ be an oriented complex triangulating $M$. Choose an orientation for $M$. Then orient the cells of the dual decomposition in such a way that adding to an orienting basis of each simplex an orienting basis of the dual cell would produce the chosen orientation of the whole manifold. Let $C_m(K)$, $C^{n-m}(K)$ be the chain group and the cochain group of complementary dimensions.

Assign to each simplex $\sigma$, considered as an element of the group $C_m(K)$, the additive functional that is equal to 1 on the dual cell $B_\sigma$ and to 0 on all the other cells. The definition of the dual cell decomposition implies that this assignment determines an isomorphism $\varphi_m$ between $C_m(K)$ and $C^{n-m}(K)$. It is easy to check that such isomorphisms are consistent with the boundary homomorphisms and coboundary homomorphisms, *i.e.* that all the diagrams

$$\begin{array}{ccc} C_m(K) & \xrightarrow{\varphi_m} & C^{n-m}(K) \\ \downarrow{\scriptstyle \partial_m} & & \downarrow{\scriptstyle \delta_{n-m}} \\ C_{m-1}(K) & \xrightarrow{\varphi_{m-1}} & C^{n-m+1}(K) \end{array}$$

are commutative. It follows that the homology groups and the cohomology groups of complementary dimensions are isomorphic. □

**Exercise 68.** Prove that if the first homology group of a closed connected three-dimensional manifold $M$ is trivial, then it is a *homology sphere*, *i.e.* that all its homology groups coincide with the homology groups $H_0 = H_3 = \mathbb{Z}$, $H_1 = H_2 = 0$ of the standard three-dimensional sphere.

Notice that if the orientation of $M$ is fixed then the isomorphisms $\varphi_m \colon C_m(K) \to C^{n-m}$ and the induced isomorphisms $\gamma_m \colon H_m(K) \to H^{n-m}(K)$ are determined uniquely. Therefore the $\cup$-product in cohomology generates a $\cap$-product in homology via the rule $a \cap b = \gamma_{k+m-n}^{-1}(\gamma_k(a) \cup \gamma_m(b))$, where $a \in H_k(M)$, $b \in H_m(M)$, and $a \cap b \in H_{k+m-n}(M)$.

The $\cap$-product too has a geometric interpretation. We limit ourselves to a brief informal description (more detailed expositions can be found in [3], [6], [9], [11]). Present $a$ as an oriented simplicial $k$-dimensional complex $A \subset M$. In other words, the chain composed of all the $k$-dimensional simplices of $A$, taken with coefficients 1, should be a cycle representing $a$. Similarly, present $b$ as an oriented simplicial complex $B \subset M$ of dimension $m$. Complexes $A$ and $B$ can be chosen so that they are *in general position*. This means that for any two open simplices $\sigma^p \in A$, $\delta^r \in B$ their intersection $\sigma^p \cap \delta^r$ must be either empty or a cell of dimension $p + r - n$. Then $A \cap B$ is a cell complex. Orient all its cells according to the following rule: if we add to an orienting basis of the cell $\sigma^p \cap \delta^r$ (it consists of $p + r - n$ vectors) $n - r$ vectors to form an orienting basis of the simplex $\sigma^p$ and then add the $n-p$ vectors in the simplex $\delta^r$, then the $n$ vectors thus obtained must determine the chosen orientation of $M$. See Figure 32. In this situation one can show that the $(p+r-n)$-dimensional cells of $A \cap B$ taken with coefficients 1 form a cycle that determines the product $a \frown b \in H_{k+m-n}(M)$.

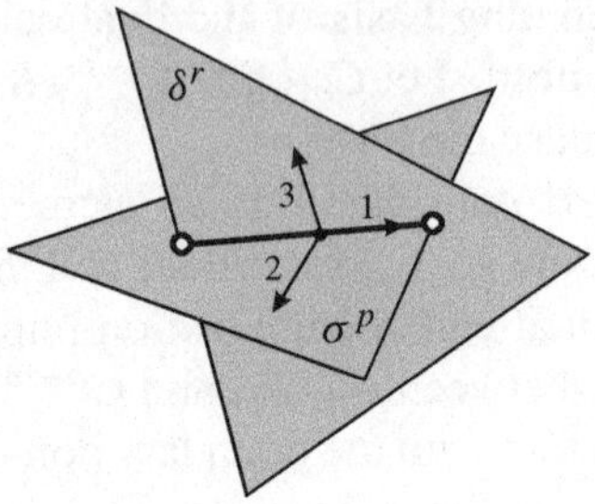


Figure 32. How to choose the correct orientation of an intersection of simplices: if we add to the collection 1 of $p+r-n$ vectors the collection 2 of $n-r$ vectors and then the collection 3 of $n-p$ vectors, then we should get an orienting basis for the manifold.

Sometimes this interpretation of the $\frown$-product in homology helps to calculate the $\smile$-product in cohomology; we replace cochains with the dual cycles, consider their intersection, and take the dual cocycle.

**Exercise 69.** Present the torus $T = S^1 \times S^1$ as the cell complex with one vertex, two one-dimensional cells $m = S^1 \times \{*\}$, $\ell = \{*\} \times S^1$, and one two-dimensional cell. Let $\mu \in H^1(T)$ be the element that is given by the cocycle that takes value 1 on $m$ and 0 on $\ell$. Similarly, let $\lambda \in H^1(T)$ correspond to the cocycle that takes value 0 on $m$ and 1 on $\ell$. Find $\mu \smile \lambda \in H^2(T) = \mathbb{Z}$.

# 2

# Elements of homotopy theory

In this section we give a rather rough account of homotopy groups. The first homotopy group (it is called fundamental) is closely related with the first homology group, since the latter is obtained from the former by taking its quotient by the commutator subgroup. Like the homology groups, the fundamental group is functorial, however, as a general rule, it carries more detailed information on the space.

## 2.1 Definition of the fundamental group

Let $X$ be a topological space. Choose an arbitrary point $x_0$, which further on will be called the base point.

**Definition.** A *loop* in a space $X$ with base point $x_0$ is an arbitrary continuous map $f\colon [0,1] \to X$ such that $f(0) = f(1) = x_0$.

**Definition.** Two loops $f$ and $g$ are called *homotopic* (notation: $f \sim g$) if there exists a continuous map $F\colon [0,1] \times I \to X$ such that $F(s,0) = f(s)$, $F(s,1) = g(s)$ for any $s \in [0,1]$, and $F(0,t) = F(1,t) = x_0$ for any $t \in I$.

In other words, compared to the general definition of homotopic maps (see page 15), we require that the homotopy be fixed at the base point.

Denoting by $f_t$, $0 \leqslant t \leqslant 1$, the restriction of $F$ to the fibre $[0,1] \times t$, we get a continuous family of maps connecting $f$ and $g$. Thus, two loops are homotopic if one can be deformed to the other by a continuous deformation.

**Exercise 70.** Which of the three loops in the annulus $S^1 \times I$ shown in Figure 33 are homotopic?

It is easy to show that the relation "homotopic to" of loops is an equivalence relation. Therefore it decomposes the set of all loops $\Omega_1(X, x_0)$ into classes of homotopic loops. The set of all such classes is denoted by $\pi_1(X, x_0)$. It forms the set of elements of the fundamental group.

Let us re-iterate: the fundamental group as a set coincides with the set of classes of homotopic loops. In other words, each element of the fundamental group is given by a loop, while two loops determine the same element of the fundamental group if and only if they are homotopic.

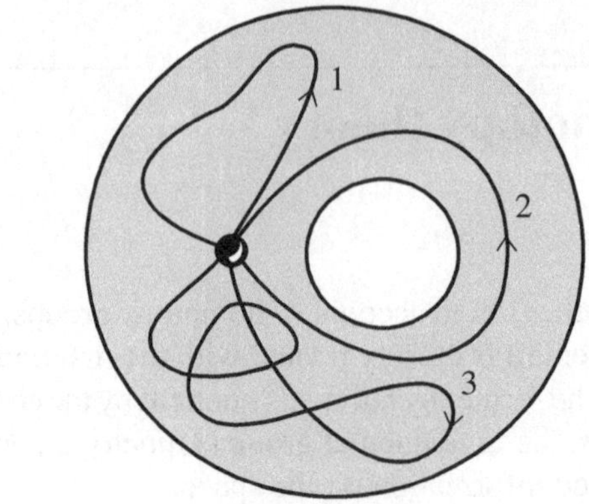


Figure 33. Two of the shown loops are homotopic to each other, but not to the third one.

We now introduce on the set $\pi_1(X, x_0)$ a binary operation, *i.e.* a rule that assigns to every two elements $\alpha$, $\beta$ of $\pi_1(X, x_0)$ a third element $\alpha\beta$. In order to do that we choose loops $f, g\colon [0,1] \to X$ representing the elements $\alpha$, $\beta$ and consider the loop $h\colon [0,1] \to X$ defined by the formula

$$h(s) = \begin{cases} f(2s), & 0 \leqslant s \leqslant 1/2, \\ g(2s-1), & 1/2 \leqslant s \leqslant 1. \end{cases}$$

The element $\alpha\beta$ is defined as the class containing the loop $h$.

The meaning of this formula is very simple: the loop $h$ is obtained from $f$ and $g$ by going first (with double velocity) along $f$ and then along $g$. The velocity is increased to ensure that the loop is defined on the segment $[0, 1]$ rather than on the segment $[0, 2]$.

**Exercise 71.** Prove that the operation is well defined, *i.e.* that the element $\alpha\beta$ does not depend on the choice of loops $f, g$ representing $\alpha$ and $\beta$.

Our main goal now is to prove that the set $\pi_1(X, x_0)$ forms a group with respect to the operation just introduced. The proof of associativity of the operation (the latter is called multiplication) consists in proving that the loops symbolically shown in Figure 34 are homotopic.

The upper segment represents the loop $k_1 = (fg)h$ defined by the formula

$$k_1(s) = \begin{cases} f(4s), & 0 \leqslant s \leqslant 1/4, \\ g(4s-1), & 1/4 \leqslant s \leqslant 1/2, \\ h(2s-1), & 1/2 \leqslant s \leqslant 1. \end{cases}$$

The lower segment represents the loop $k_2 = f(gh)$.

**Exercise 72.** Prove that the loops $k_1$ and $k_2$ are homotopic (write down a formula for the homotopy).

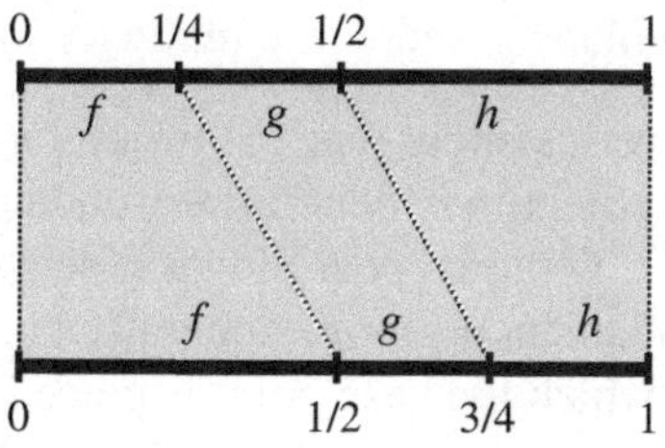


Figure 34. The loops $(fg)h$ and $f(gh)$ only differ in their parametrization.

In order to prove the existence of the identity element we should describe it and prove that it is indeed the identity.

**Exercise 73.** Show that the constant loop (which maps the whole segment to the base point) defines the identity element in the set $\pi_1(X, x_0)$ with respect to multiplication.

The element inverse to a given one is obtained by taking a loop representing that element and going around it in the reverse direction.

**Exercise 74.** Prove that going around a loop in two opposing directions defines elements of $\pi_1(X, x_0)$ inverse to each other with respect to multiplication.

Thus, the set $\pi_1(X, x_0)$ does form a group with respect to the defined operation.

**Definition.** The group $\pi_1(X, x_0)$ is called the *fundamental group* of the given space $X$ with base point $x_0$.

## 2.2 Independence of the choice of the base point

**Definition.** A topological space $X$ is called *pathwise connected* if any two of its points can be joined by a continuous path. This means that for any two points $x_1, x_2 \in X$ there must be a continuous map $s\colon [0,1] \to X$ such that $s(0) = x_1$ and $s(1) = x_2$.

**Theorem 32.** *If a space $X$ is pathwise connected, then for any two points $x_1, x_2 \in X$ the groups $\pi_1(X, x_1)$ and $\pi_1(X, x_2)$ are isomorphic.*

To prove this we choose a path $s$ that joins $x_1$ to $x_2$ and define a map $\varphi\colon \pi_1(X, x_1) \to \pi_1(X, x_2)$ as follows: $\varphi(\alpha) = [s^{-1}\widetilde{\alpha}s]$, where $\widetilde{\alpha}$ is a loop representing the element $\alpha$ and $[g]$ denotes the class containing the loop $g$. The inverse map $\psi\colon \pi_1(X, x_2) \to \pi_1(X, x_1)$ is defined in a similar way: $\psi(\beta) = [s\widetilde{\beta}s^{-1}]$.

**Exercise 75.** Prove that the maps $\varphi$ and $\psi$ are well defined, *i.e.* do not depend on the choice of loops representing given elements of the fundamental groups.

**Exercise 76.** Prove that the maps $\varphi$ and $\psi$ are mutually inverse isomorphisms.

The results of these exercises show that the groups $\pi_1(X, x_1)$ and $\pi_1(X, x_2)$ are isomorphic. Therefore we sometimes denote the fundamental group of a pathwise connected space $X$ as $\pi_1(X)$, without specifying a base point. For a space that is not pathwise connected the fundamental group, generally speaking, does depend on the choice of the component in which the base point is placed.

Suppose that we have a continuous map $f\colon X \to Y$ between two spaces that preserves the base points (*i.e.* takes the base point $x_0$ of the space $X$ to the base point $y_0$ of the space $Y$). Define the induced map $f_*\colon \pi_1(X, x_0) \to \pi_1(Y, y_0)$ in the following way: $f_*(\alpha) = [f\widetilde{\alpha}]$ (recall that $\widetilde{\alpha}$ denotes any loop representing $\alpha$ and $[f\widetilde{\alpha}]$ is the element of the fundamental group defined by the loop $f\widetilde{\alpha}$). One can think of this as follows: to each loop in $X$ we assign the loop in $Y$ that is the image of the first loop under the map $f$.

It is easy to check that the map $f_*$ is well defined (*i.e.* that $f_*(\alpha)$ does not depend on the choice of the representing loop $\widetilde{\alpha}$) and that the following properties hold:

1. The map $f_*$ is a homomorphism (this homomorphism is called the *induced* homomorphism).

2. The identity map induces the identity homomorphism of the groups.

3. A composition of maps between spaces induces the composition of the induced homomorphisms of the respective groups.

Having these properties means that assigning to each space with a marked base point its fundamental group is a functor from the category of pointed (*i.e.* endowed with a marked base point) topological spaces to the category of groups.

Let us give several examples. It is easy to see that the fundamental group of the space that consists of a single point is trivial. Let us prove that the fundamental group of the segment $[0, 1]$ is also trivial. Indeed, if $f\colon [0, 1] \to [0, 1]$ is a loop, then the formula $F(x, t) = t + (1 - t)f(x)$ defines a homotopy between this loop and the constant loop (we assume that the base point is 1). This means that $\pi_1([0, 1])$ consists of just one element.

**Exercise 77.** Prove that the fundamental group of any convex subset of a Euclidean space is trivial.

Next we show that the fundamental group of the circle is isomorphic to the group of integers $\mathbb{Z}$. Let $S^1$ be the unit circle on the plane with the center at the origin. Choose $(1, 0)$ as the base point. Let $f\colon [0, 1] \to S^1$ be a loop. When the parameter $t$ runs from 0 to 1, the point $f(t)$ travels along the circle and at the final moment $t = 1$ returns to the base point. Therefore the total number of times it winds around the circle is an integer. It is not difficult to see that this number is unchanged by deformations (homotopies) of the loop. Indeed, under a continuous deformation of the loop the number of windings should be changing continuously as well, and since it is always an integer, it must remain constant.

**Exercise 78.** Prove that assigning to each loop the number of times it winds around the circle determines an isomorphism between $\pi_1(S^1)$ and $\mathbb{Z}$.

## 2.3 Presentations of groups

In order to learn how to calculate the fundamental groups of more complicated spaces, we need to learn how to present groups. One of the most common and efficient ways of describing a group is to present it by means of generators and relations.

Let $A$ be an arbitrary finite set. We call it an alphabet and its elements letters. A word in the alphabet $A$ is an arbitrary finite sequence of symbols of form $a, a^{-1}$, where $a \in A$. The *length of a word* is the number of letters involved in it. Here are a few examples of words in an alphabet with two letters: $a$, $b$, $aa^{-1}ba$, $ababbab^{-1}$. The lengths of these words are 1, 1, 4, and 7.

**Exercise 79.** How many different words of length 2 can one write using an alphabet with two letters?

It is convenient to use the following conventions: the symbol $a^n$, where $n$ is a natural number, denotes the word $\underbrace{aa\dots a}_{n}$, the symbol $a^{-n}$ is a short representation of the word $\underbrace{a^{-1}a^{-1}\dots a^{-1}}_{n}$, and the symbol 1 denotes the empty word.

Let $R_1, R_2, \dots, R_n$ be some words in an alphabet $A$ that consists of letters $a_1, a_2, \dots, a_m$. We describe a construction that to the given data (the letters $a_i$, which are called *generators*, and the words $R_j$, which are called *defining relations*) assigns a certain group $G$ denoted as

$$\langle a_1, a_2, \dots, a_m \mid R_1, R_2, \dots, R_n \rangle.$$

First we introduce on the set $W$ of all words in the alphabet $A$ the following equivalence relation: a word $w_1$ is *equivalent* to a word $w_2$ if one can pass from $w_1$ to $w_2$ by a finite sequence of moves of the form I, I′ and II:

I. Insert or delete a pair of form $a_i a_i^{-1}$.

I′. Insert or delete a pair of form $a_i^{-1} a_i$.

II. Find and delete a subword $R_j$, or insert such a subword.

**Exercise 80.** Prove that this is indeed an equivalence relation.

Relations are sometimes written not in the form of words but as equalities of type $R = 1$ or even of type $R = Q$, where $R$, $Q$ are some words in the generators. In the second case move II consists in finding a sub-word $R$ and replacing it by the sub-word $Q$ (or vice versa). Of course, the relation $R = Q$ basically coincides with the relation $RQ^{-1} = 1$.

**Exercise 81.** Let $A = \{a, b\}$, and let the set of relations consist of two relations, $abab^{-1} = 1$ and $b^2a^{-2} = 1$. Prove that the word $a^4$ is equivalent to the empty word 1.

Next we introduce on the set $W/\sim$ of the equivalence classes an operation of multiplication, setting $\alpha\beta = [\widetilde{\alpha}\widetilde{\beta}]$. Here $\widetilde{\alpha}$ denotes a word representing the class $\alpha$, and $[\widetilde{\alpha}\widetilde{\beta}]$ denotes the class containing the word $\widetilde{\alpha}\widetilde{\beta}$. In other words, to multiply two classes we should choose a representative for each one of them, append the word representing the second class to the one representing the first class from the right, and take the class containing the resulting word.

The correctness of the above definition (*i.e.*, its independence on the choice of representing words) is proven in the usual way.

**Exercise 82.** Prove that the set $G = W/\sim$ is a group with respect to the above operation.

Taking some liberty, we can think of $G$ as consisting of all words in the alphabet $A$, with two words defining the same element if and only if they are equivalent.

**Exercise 83.** Prove that $G$ remains unchanged if one of the relations $R_i$ is replaced by its conjugate, *i.e.* by a relation of the form $wR_iw^{-1}$, where $w$ is some word in the alphabet $A$.

Let $w$ be some word in the alphabet $A$. We say that the relation $w = 1$ is *deducible from relations $R_i$, $1 \leqslant i \leqslant n$,* if $w$ determines the identity element of the group $\langle a_1, \dots, a_m \mid R_1, \dots, R_n \rangle$. We remark that an actual process of deducing a new relation from given ones requires some ingenuity and luck. There is no general algorithm.

**Exercise 84.** Prove that the relation $z^6 = 1$ is deducible in the group

$$\langle x, z \mid xz^2xz^{-1}, zx^2zx^{-1} \rangle.$$

**Exercise 85.** Prove that the group remains unchanged by the following operations applied to its presentation:

I. Adding a new relation, deducible from the given relations of the group.

I′. Removing a relation that is deducible from the remaining ones.

II. Adding a new generator $a$ and a new relation of form $a = w$, where the word $w$ does not contain the generator $a$.

II′. Removing a generator $a$ and a relation $a = w$ provided that $a$ occurs neither in $w$ nor in any other relations.

It is easy to prove that if the presentations $\langle a_1, \dots, a_m \mid R_1, \dots, R_n \rangle$ and $\langle b_1, \dots, b_k \mid Q_1, \dots, Q_l \rangle$ define isomorphic groups, then one can pass from one of them to the other by a sequence of moves I, I′, II, II′. However, determining a specific sequence of moves can be very difficult. Moreover, it has been proven that there is no general algorithm to tell whether two given groups are isomorphic.

We give two examples of recognizing a group given by generators and relations. To recognize a group means for us to show that it is isomorphic to one of the known groups. Recall again that sometimes relations are written in the form $w_1 = w_2$ rather than in the form $w_1 w_2^{-1} = 1$. For instance, the relation $xz^2xz^{-1} = 1$ can be written as $xz^2x = z$ and the two relations $a^2(ab)^{-3} = 1$ and $a^2b^{-3} = 1$, as $a^2 = (ab)^3 = b^3$.

**Example.** Identify the group

$$\langle a, b, c, d, e \mid d = e^2, bda = 1, ab^{-1}c = 1, ac^{-1}b^{-1} = 1, de = c \rangle.$$

*Solution.* We use the first relation $d = e^2$ to remove the generator $d$. We get the presentation

$$\langle a, b, c, e \mid be^2a = 1, ab^{-1}c = 1, ac^{-1}b^{-1} = 1, e^3 = c \rangle.$$

Using the last relation $c = e^3$, we remove $c$:

$$\langle a, b, e \mid be^2a = 1, ab^{-1}e^3 = 1, ae^{-3}b^{-1} = 1 \rangle.$$

Using the relation $b = e^3a$, which is equivalent to $ab^{-1}e^3 = 1$, we remove $b$:

$$\langle a, e \mid e^3ae^2a = 1, ae^{-3}a^{-1}e^{-3} = 1 \rangle, \text{ or } \langle a, e \mid e^3ae^2a = 1, e^3ae^3 = a \rangle.$$

Rewriting the first relation as $e^3ae^3e^{-1}a = 1$ and replacing $e^3ae^3$ by $a$, we get the presentation

$$\langle a, e \mid ae^{-1}a = 1, e^3ae^3 = a \rangle, \quad \text{or} \quad \langle a, e \mid e = a^2, e^3ae^3 = a \rangle.$$

Finally, removing $e$ with the help of the first relation, we get the presentation $\langle a \mid a^6aa^6 = a \rangle$, *i.e.* $\langle a \mid a^{12} = 1 \rangle$. This is the cyclic group of order 12.

**Example.** Identify the group $G = \langle a, b \mid a^3 = b^2 = 1, a^2b = ba \rangle$.

*Solution.* We take an arbitrary word $w = a^{\alpha_1}b^{\beta_1} \dots a^{\alpha_n}b^{\beta_n}$ in the alphabet $a, b$ and try to simplify it using the given relations. The first two relations allow us to make all the $\beta_i$ equal to 1 and all the $\alpha_i$ equal to 1 or 2. The third relation allows us to move all the occurences of $b$ to the end of the word $w$. This of course changes the powers at the letters $a$. Thus, $w$ is equivalent to a word $a^\alpha b^\beta$, where $\alpha$ can take value 0, 1, or 2, and $\beta$ can take value 0 or 1. Thus, the group contains at most six elements. The only non-Abelian group with six elements is the symmetric group $S_3$, which consists of permutations of numbers 1, 2, 3 and where the product of two permutations is obtained by applying the second permutation and then the first one. It is easy to check that assigning to $a$ the permutation (231) and to $b$ the permutation (213) preserves the relations $a^3 = b^2 = 1$, $a^2b = ba$, *i.e.* these equalities are transformed into correct equalities in $S_3$. Therefore this assignment can be extended to a homomorphism $\varphi\colon G \to S_3$. This homomorphism is surjective, since the permutations (231) and (213) generate $S_3$. The injectivity follows from the surjectivity and the fact that $S_3$ contains exactly six elements, while $G$ contains at most six. Thus, $G$ is isomorphic $S_3$.

**Exercise 86.** Prove that the group $G = \langle x, y \mid xy^2x = y, yx^2y = x\rangle$ is isomorphic to the binary tetrahedral group $\langle a, b \mid a^2 = (ab)^3 = b^3, a^4 = 1\rangle$.

Adding to the relations of an arbitrary presentation the commutativity relations yields a new group, which is Abelian. This operation is called taking the Abelianization. The Abelianizations of isomorphic groups are isomorphic, while the reverse is not true.

**Exercise 87.** Prove that the groups

$$\langle a, b \mid a^2b^3a^{-1}b = 1, abab^2 = 1\rangle$$

and

$$\langle x, y \mid x^{-1}y^3xy = 1, x^5yx^{-2}y^{-3} = 1\rangle$$

are distinct.

## 2.4 Calculation of fundamental groups

To calculate a fundamental group means to write down a presentation of it. The van Kampen theorem allows us to calculate the fundamental group of the union of two spaces $X \cup Y$, when the fundamental groups of the spaces $X$, $Y$, and $Z = X \cap Y$ are known.

**Theorem 33.** *Suppose that linearly connected cell complexes $X$, $Y$ intersect along their common connected subcomplex $Z = X \cap Y$. Then a presentation of the group $\pi_1(X \cup Y)$ can be obtained as follows:*

1. *Write down some generators of $\pi_1(X)$ and $\pi_1(Y)$.*
2. *Write down relations for $\pi_1(X)$ and $\pi_1(Y)$.*
3. *Write down one more series of relations, one for each generator $c$ of the group $\pi_1(Z)$. This relation has the form $\varphi_1(c) = \varphi_2(c)$, where $\varphi_1(c)$ is an expression of $c$ in the generators of $\pi_1(X)$ and $\varphi_2(c)$ is its expression in the generators of $\pi_1(Y)$.*

A proof of the van Kampen theorem can be found, for instance, in [6].

**Exercise 88.** Calculate the fundamental group of the figure-eight (the wedge of two circles).

**Exercise 89.** Calculate the fundamental group of the projective plane.

As an application of the van Kampen theorem we prove that the trefoil knot cannot be untied, a fact that is probably known to everyone from manipulating his or her own shoelaces.

**Theorem 34.** *The closure of the complement in $S^3$ of the standard solid torus is again homeomorphic to the solid torus.*

The idea of the proof is shown in Figure 35. Rotating the disc $D$ along the axis $l$ yields a solid torus $V_1$. The closure $V_2$ of its complement consists of discs parameterized by points of $l$ (which, together with $\infty$, form a circle). Moreover, each point of $l$ is intersected by exactly one such disc, the one obtained by rotating the "solenoid" arc around $l$. Therefore $V_2$ is also a solid torus. Notice that the meridian of the interior torus coincides with the parallel of the exterior one, and vice versa.

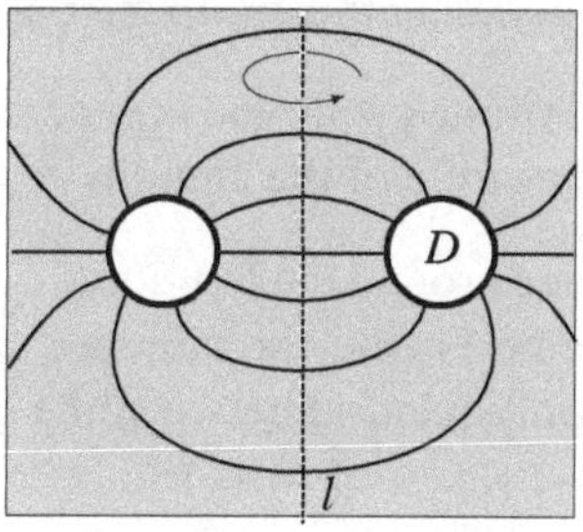


Figure 35. Decomposition of $S^3$ into two solid tori.

**Theorem 35.** *The trefoil knot (see Figure 36) cannot be untied (in rigorous terms, there is no isotopy of $\mathbb{R}^3$ that would take the trefoil $K$ to the standard circle $S^1 \subset \mathbb{R}^3$).*

*Proof.* Recall that the *complement* $C_K$ of an arbitrary knot $K \subset S^3$ is defined as $C_K = S^3 \setminus \operatorname{Int} N(K)$: we first replace $K$ with its tubular neighbourhood $N(K)$, which is a knotted solid torus, and then remove its interior. The difference from its complement in the general sense, $S^3 \setminus K$, is that $C_K$ is a compact space (which is convenient). If $K$ is trivial, then the fundamental group of the complement (which is homeomorphic to the solid torus) is the infinite cyclic group. Therefore it is Abelian. Thus, to prove that the trefoil is non-trivial it is sufficient to find the fundamental group of its complement and to prove that it is not Abelian.

Let $K$ be the trefoil. We place it on the common boundary $T = V_1 \cap V_2$ of the solid tori $V_1, V_2 \subset S^3$, as it is shown in Figure 36. Then $V_i' = V_i \cap C_K$, $i = 1, 2$,

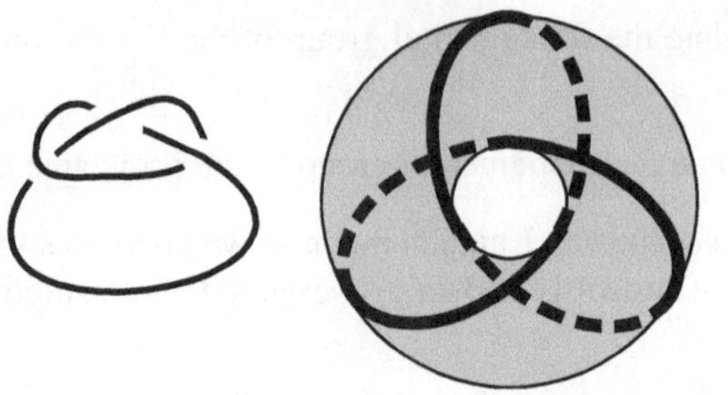

Figure 36. The trefoil and its position on the boundary of the standard solid torus.

are also solid tori that are obtained from $V_1$, $V_2$ by cutting out small tunnels along $K$. They intersect along the annulus $A = T \cap C_K$ and together yield $C_K$.

To calculate $\pi_1(C_K)$ we apply the van Kampen theorem.

1. The groups $\pi_1(V_1')$ and $\pi_1(V_2')$ are infinite cyclic groups, whose generators $v_i \in \pi_1(V_i')$, $i = 1, 2$, can be chosen to be elements given by the core circles of the solid tori. Hence $v_1$, $v_2$ generate the group $\pi_1(C_K)$.

2. The group $\pi_1(A)$ is also infinite cyclic and is generated by any element $a \in \pi_1(A)$ corresponding to the core circle of the annulus $A$.

3. Since the trefoil goes twice around the longitude of the torus $T$ and thrice around its meridian (and the same is true for the core circle of $A$), we have $l = v_1^2$ and $l = v_2^3$. Hence the only relation between the generators $v_1$, $v_2$ has the form $v_1^2 = v_2^3$ and $\pi_1(C_K) = \langle v_1, v_2 \mid v_1^2 = v_2^3 \rangle$.

The presentation of $\pi_1(C_K)$ just obtained allows us to conclude that it is non-Abelian. For this it suffices to see that the permutation group $S_3 = \langle a, b \mid a^3 = b^2 = 1, a^2b = ba \rangle$ of degree 3 (see the example on page 67) is its epimorphic image. The desired epimorphism can be given by the rule $v_1 \to a$, $v_2 \to b$. □

**Example.** Suppose that the intersection $Z = X \cap Y$ of two connected cell complexes consists of exactly one point. In this case we say that the complex $X \cup Y$ is the *wedge* of complexes $X$, $Y$. Then the group $\pi_1(X \cup Y)$ has presentation

$$\pi_1(X \cup Y) = \langle a_1, \dots, a_m, b_1, \dots, b_p \mid R_1, \dots, R_n, Q_1, \dots, Q_q \rangle,$$

where $\langle a_1, \dots, a_m \mid R_1, \dots, R_n \rangle$ and $\langle b_1, \dots, b_p \mid Q_1, \dots, Q_q \rangle$ are presentations of $\pi_1(X)$ and $\pi_1(Y)$ respectively. This means that the group $\pi_1(X \cup Y)$ is the free product of $\pi_1(X)$ and $\pi_1(Y)$.

In particular, if in the above example the complex $Y$ is a circle, then a presentation of $\pi_1(X \cup Y)$ can be obtained from some presentation of $\pi_1(X)$ by adding one new generator. The relations remain the same.

**Example.** Suppose that a cell complex $X_1$ is obtained from a connected cell complex $X$ by gluing to it one two-dimensional cell along some map of its boundary curve. Then a presentation of $\pi_1(X_1)$ can be obtained from a presentation of $\pi_1(X)$ by adding exactly one relation that shows how the boundary curve of the added cell is expressed through the generators of $\pi_1(X)$.

These two examples bring us to a simple method of finding a presentation of the fundamental group of an arbitrary cell complex. This method is rather close to the method of calculating the first homology group described on page 45.

**Calculating the fundamental group of a cell complex.** Let $X$ be a connected cell complex. Its fundamental group $\pi_1(X)$ is fully determined by its two-dimensional skeleton $X^{(2)}$, which is obtained from the one-dimensional skeleton $\Gamma = X^{(1)}$ by gluing to it several two-dimensional cells.

*Step* 1. Choose a *maximal tree* in $\Gamma$, *i.e.* a subgraph of $\Gamma$ which does not contain cycles and which passes through all the vertices of $\Gamma$. Orient the remaining edges and label them by letters. These letters are the generators of $\pi_1(X)$ in our presentation.

*Step* 2. For each two-dimensional cell we write a word that shows how the boundary circle of this cell passes along the labelled edges. These words form a set of relations of $\pi_1(X)$.

**Example.** Suppose that a cell complex $K$ is obtained from the graph shown in Figure 37 by gluing two cells of dimension 2 along the maps of the circle indicated there. Application of the algorithm shows that

$$\pi(K) = \langle a, b, c \mid ab^{-1}, cb^{-1} \rangle = \mathbb{Z}.$$

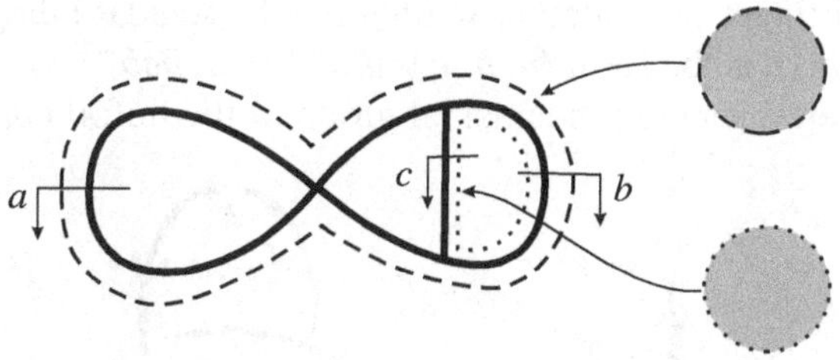


Figure 37. A simple example of calculating the fundamental group.

**Exercise 90.** Present the Klein bottle as a cell complex and find its fundamental group.

**Exercise 91.** Find the fundamental group of the handlebody of genus 2.

**Exercise 92.** Prove that for any group $G$ with finitely many generators and relations there is a space $X$ whose fundamental group is $G$.

The resemblance between the methods of calculating the first homology group and the fundamental group is not accidental. It is easy to show that the first homology group is isomorphic to the Abelianization of the fundamental group. Indeed, when computing the first homology group we follow essentially the same procedure but ignore the order in which the boundary curves pass along edges. This amounts to taking the Abelianization.

## 2.5 Wirtinger's presentation

The above method of calculating the fundamental group of the complement of the trefoil works also for other *torus knots* (*i.e.* knots that can be placed on the boundary of the standard solid torus in $S^3$). In this section we describe another method, which can be applied to any knot. The presentations obtained with its help are called *Wirtinger presentations*.

Consider a knot $K$ given by its projection to $\mathbb{R}^2$, which, as customary in knot theory, is disconnected in the double points to indicate which portion of the diagram passes above the other one. Such a disconnected projection is called a *knot diagram*. From the topological point of view it is a collection of disjoint arcs. We orient these arcs in such a way that together they form one of the two possible orientations of the knot, and we denote them by letters. These letters are the generators in Wirtinger's presentation. Then for each double point we write the relation $xyx^{-1} = z$, where $x$ denotes the arc going above, $y$, the arc meeting it from the right, and $z$, the arc meeting it from the left. See Figure 38, left. The orientations of $y$ and $z$ do not play any rôle here. The obtained presentation is called a Wirtinger presentation.

**Example.** The Wirtinger presentation written for the standard diagram of the trefoil (see Figure 38, right) has the form $\langle a, b, c \mid aba^{-1} = c, bcb^{-1} = a, cac^{-1} = b\rangle$. Let us show that it determines the fundamental group of the trefoil complement that was

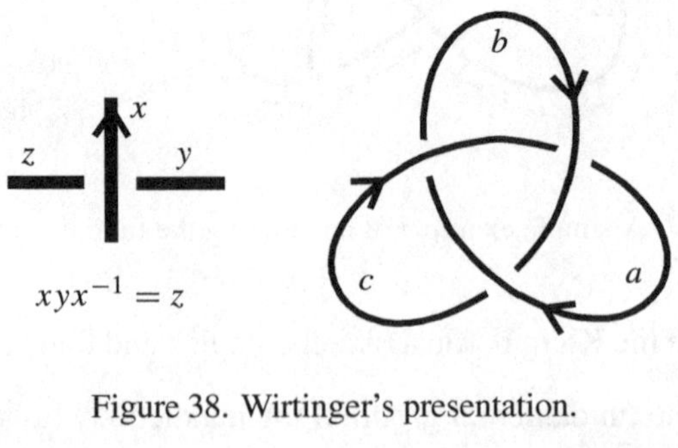


Figure 38. Wirtinger's presentation.

obtained in the proof of Theorem 35. Each of the three relations in this presentation can be deduced from the other two. Therefore one relation can be discarded. This is a general fact, which holds for all Wirtinger presentations. We discard the third relation and use the second one to express the generator $c$ through $a$ and $b$. We get the presentation $\langle a, b \mid baba^{-1}b^{-1} = a\rangle$, which is equivalent to $\langle a, b \mid aba = bab\rangle$. We now use the replacement $v_1 = aba$, $v_2 = ab$ to transform the latter presentation to the presentation $\langle v_1, v_2 \mid v_1^2 = v_2^3\rangle$ obtained in the proof of Theorem 35.

Thus the Wirtinger presentation written for the trefoil's standard diagram determines the fundamental group of the trefoil complement. This turns out to be a general fact.

**Theorem 36.** *For any diagram of a knot $K$ the Wirtinger presentation corresponding to that diagram defines the fundamental group of its complement.*

*Proof.* Denote by $K'$ the knot whose diagram is obtained from the given diagram of $K$ by changing the types of all double points, see Figure 39 (middle). Then $K'$ is obtained from $K$ by the mirror reflection along the plane. Since the fundamental groups $\pi_1(C_K)$, $\pi_1(C_{K'})$ of their complements are isomorphic, it is sufficient to show that the Wirtinger presentation written according to the given diagram of $K$ defines the group $\pi_1(C_{K'})$.

Let us draw the given diagram of $K$ on a sphere $S$ (the boundary of a ball $B^3 \subset S^3$) and glue to $B^3$ strips (tunnels) along the arcs of the diagram, one strip for each arc. Each strip is in fact a rectangle glued to the ball along a pair of opposite sides, therefore the gluing results in adding a new generator to the fundamental group. Thus, the fundamental group of the ball with the attached strips is free, with free generators corresponding to the arcs of the diagram.

Now we glue a two-dimensional cell in the neighbourhood of each double point, as is shown in Figure 39, right: the cell joins the exits of two tunnels and passes twice

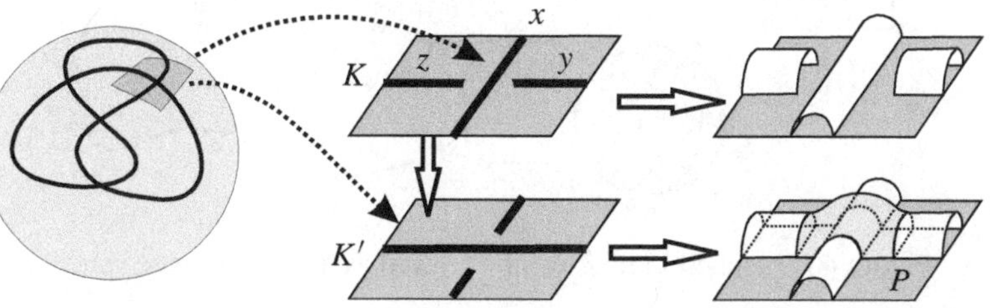


Figure 39. Gluing strips and cells to a ball models both the space $C_{K'}$ and the relevant Wirtinger presentation of $\pi_1(C_K)$.

along the roof of the third tunnel. Adding such a cell yields the relation $xyx^{-1}z^{-1} = 1$, which essentially coincides with a Wirtinger relation $xyx^{-1} = z$. As a result, we get a polyhedron $P$ that possesses two properties:

1. The group $\pi_1(P)$ admits a presentation that coincides with the Wirtinger presentation corresponding to the given projection of $K$.

2. The groups $\pi_1(P)$ and $\pi_1(C_{K'})$ are isomorphic.

The first property is true by construction, the second one easily follows from the fact that the complement of $P$ in $S^3$ consists of the tubular neighbourhood of $K'$ and two open three-dimensional balls. Since $\pi_1(C_K) = \pi_1(C_{K'})$, the Wirtinger presentation assigned to the given diagram of $K$ indeed defines the fundamental group of its complement. □

## 2.6 The higher homotopy groups

The higher homotopy groups $\pi_n(X, x_0)$ are defined in the same manner as the fundamental group. The difference is that instead of loops we consider *spheroids*.

**Definition.** A *spheroid of dimension n* in a space $X$ with base point $x_0$ is a continuous map $f: I^n \to X$ such that $f(\partial I^n) = x_0$.

We have called a map from the cube a spheroid, because the boundary of the cube is mapped to a point and therefore the map filters through a map to the sphere. See Figure 40.

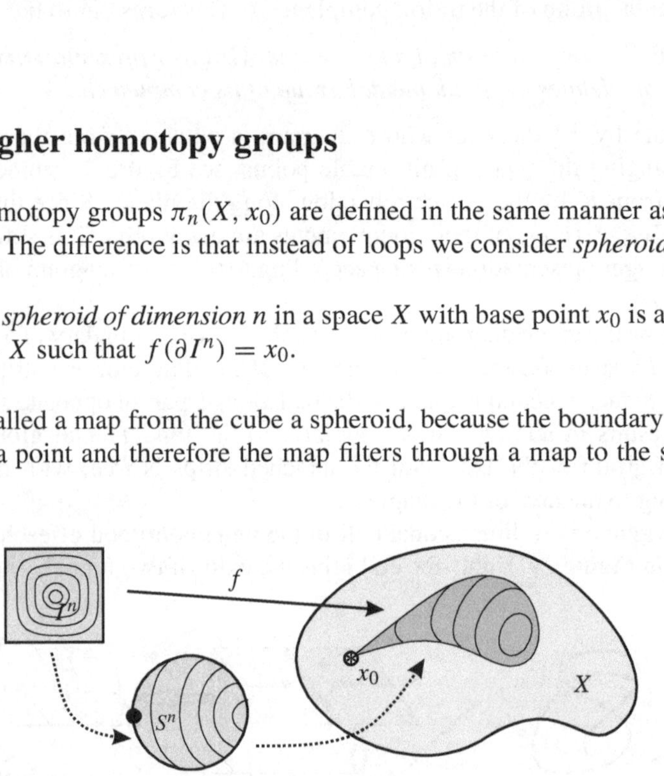


Figure 40. A spheroid in $X$ can be viewed as a map from the sphere.

As in the case of loops, a homotopy between spheroids should be fixed, *i.e.* the boundary of the cube should always be mapped to the base point. We introduce on the set $\pi_n(X, x_0)$ of classes of homotopic spheroids a binary operation (certainly, using representatives). As will be seen below, for $n > 1$ this operation is commutative. Therefore it is called addition.

Let $f_1, f_2: I^n \to X$ be two spheroids representing two given elements $\alpha_1, \alpha_2 \in \pi_n(X, x_0)$. Choose inside the cube $I^n$ two copies $I_1^n, I_2^n$ of the standard cube $I^n$ that are obtained by homothethies $h_1: I^n \to I_1^n$, $h_2: I^n \to I_2^n$. The cubes $I_1^n, I_2^n$ should not have common points. Then a spheroid $g: I^n \to X$ representing the sum $\alpha_1 + \alpha_2$

can be defined as follows (see Figure 41):

$$g(x) = \begin{cases} f_1 h_1^{-1}(x), & \text{if } x \in I_1^n, \\ f_2 h_2^{-1}(x), & \text{if } x \in I_2^n, \\ x_0, & \text{if } x \notin I_1^n \cup I_2^n. \end{cases}$$

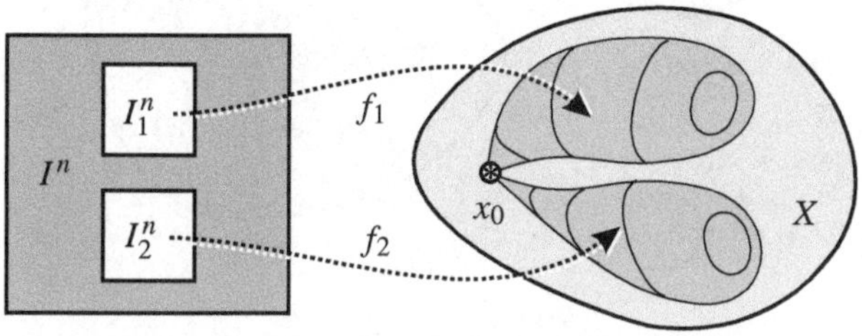


Figure 41. When the spheroids are summed up, the shadowed region is mapped to the base point.

A homotopy between spheroids $f+g$ and $g+f$ that proves that the above operation is commutative is shown in Figure 42. At any moment of this homotopy the image of the sum of spheroids remains the same, however the map of the cube $I^n$ in $X$ is continuously deformed. This happens because the smaller cubes $I_1^n$, $I_2^n$ exchange places inside $I^n$ without touching each other.

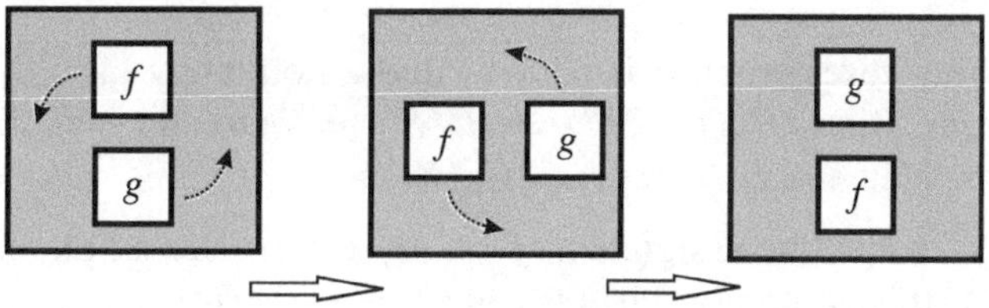


Figure 42. Addition of spheroids is commutative.

**Exercise 93.** Prove that the set $\pi_n(X, x_0)$ forms a group with respect to the above operation.

Calculating the higher homotopy groups, despite their being Abelian, is a very difficult task. There is no general way of selecting generators. Moreover, the homotopy groups of compact polyhedra are often infinitely generated. One possible explanation of this phenomenon is that the fundamental group $\pi_1(X, x_0)$ acts on all the higher homotopy groups $\pi_n(X, x_0)$. This means that the groups $\pi_n(X, x_0)$ are modules over $\pi_1(X, x_0)$.

The simplest way to describe the action of the fundamental group is the following. First we compress a half of the sphere $S^n$ to a segment and map the result onto the wedge of another sphere and a segment. Then we map this wedge to $X$, as it is specified by the given elements $\alpha \in \pi_1(X, x_0)$ and $\beta \in \pi_n(X, x_0)$. The obtained map from $S^n$ to $X$ determines the result $\alpha(\beta)$ of the action of the element $\alpha$ on the element $\beta$, see Figure 43.

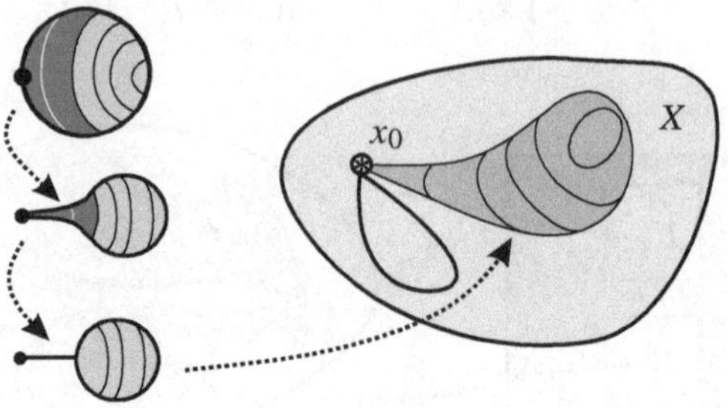


Figure 43. The action of the fundamental group on $\pi_n(X)$.

**Exercise 94.** Prove that for any element $\alpha \in \pi_1(X, x_0)$ the left shift $\tau_\alpha : \pi_n(X, x_0) \to \pi_n(X, x_0)$ is an isomorphism.

If in the above construction instead of the loop we take an arbitrary path joining the base point $x_0$ with another point $x_1 \in X$, then we get an isomorphism from the group $\pi_n(X, x_0)$ to the group $\pi_n(X, x_1)$. Therefore for a pathwise connected space $X$ the group $\pi_n(X, x_0)$ does not depend on the choice of the base point, and we can write simply $\pi_n(X)$.

One of the few cases when it is easy to find a homotopy group is given by the Hurewicz theorem. Since $H_n(I^n, \partial I^n) = H_n(S^n) = \mathbb{Z}$, to any spheroid $f : I^n \to X$ we can assign the element $\alpha_f = f_*(1)$ of $H_n(X)$.

**Exercise 95.** Prove that the assignment $f \mapsto \alpha_f$ determines a homomorphism from $\pi_n(X)$ to $H_n(X)$ (this homomorphism is called the *Hurewicz homomorphism*).

**Theorem 37.** *Suppose that a simply-connected polyhedron $X$ is such that for all $k$, $1 \leqslant k < n$, the groups $H_k(X)$ are trivial. Then the Hurewicz homomorphism $\pi_n(X) \to H_n(X)$ is an isomorphism.*

The proof can be found in [11], [4], [10].

**Exercise 96.** Prove that $\pi_n(X \times Y) = \pi_n(X) \oplus \pi_n(Y)$.

## 2.7 Bundles and exact sequences

**Definition.** A *bundle* is an arbitrary continuous map $p : E \to B$ from one topological space *onto* another one. Here $E$, $B$, and $p$ are called, respectively, the *total space*, the *base space*, and the *projection map* of the bundle.

The idea of replacing the words "map onto" with the word "bundle" is to emphasize in what way the space $E$ is decomposed into *fibres*, the pre-images of points under the projection map $p$. An example of a bundle is shown in Figure 44. In this example the total space is decomposed into fibres of four types: a point, a segment, two segments, and a point and a segment.

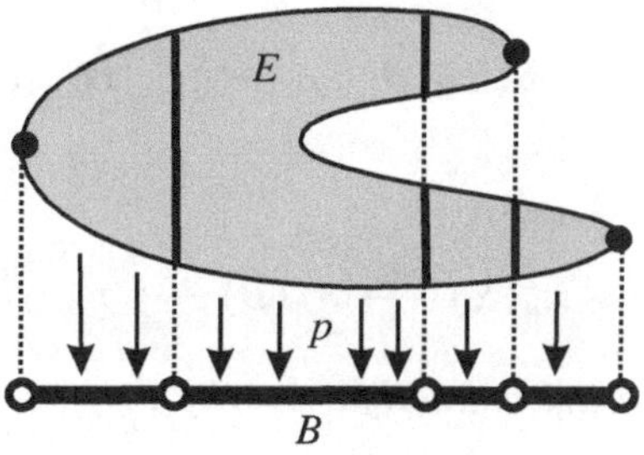


Figure 44. Decomposition of the total space into fibres.

Bundles form a category, where the morphisms are maps between total spaces that take fibres to fibres. Therefore two bundles $p_1\colon E_1 \to B_1$ and $p_2\colon E_2 \to B_2$ are isomorphic if there exist homeomorphisms $f\colon E_1 \to E_2$, $g\colon B_1 \to B_2$ such that $p_2 f = g p_1$. The bundle $p\colon B \times F \to F$ whose projection map is given by the rule $p(b, f) = b$, and any bundle isomorphic to it is called *trivial*.

**Definition.** A bundle $p\colon E \to B$ is called *locally trivial* if for any point $x \in B$ there is an open neighbourhood $U \ni x$ such that the restriction $p_U\colon E_U \to U$ of $p$ to the pre-image $E_U = p^{-1}(U) \subset E$ is a trivial bundle.

In other words, a bundle is locally trivial if locally (*i.e.* on the pre-image of some neighbourhood of each point of the base) it is a direct product. Obviously, all fibres of a locally trivial bundle with connected base are homeomorphic.

**Definition.** We say that a map $\tilde{f}\colon P \to E$ from a polyhedron $P$ to the total space of a bundle $p\colon E \to B$ is a *lift* of a map $f\colon P \to B$ if $p\tilde{f} = f$. A bundle possesses the *homotopy lifting property* if for any polyhedron $P$, for any homotopy $f_t\colon P \to B$, $0 \leqslant t \leqslant 1$, and for any lift of the initial map $f_0$ there exists a consistent lift $\tilde{f}_t\colon P \to E$ of the whole homotopy.

**Exercise 97.** Prove that the trivial bundle $p\colon B \times F \to B$ has the following relative homotopy lifting property. Suppose that we are given not only the homotopy $f_t\colon P \to B$ and its lift $\tilde{f}_0$ at the initial moment but also a consistent lift $\tilde{g}_t\colon Q \to E$ of the restriction of the homotopy $f_t$ to some subpolyhedron $Q \subset P$. Then the lift $g_t$ can be extended to some lift $\tilde{f}_t\colon P \to E$.

Decomposing the given homotopy $f_t\colon P \to B$ into a composition of small deformations and using the results of Exercise 97, we can prove that any locally trivial bundle possesses the homotopy lifting property.

Let $p\colon E \to B$ be a locally trivial bundle with connected total space. Choose base points $x_0 \in B$ and $y_0 \in B$ such that $p(x_0) = y_0$. Denote by $F$ the fibre that contains the point $x_0$. Then the embedding $i\colon F \to E$ and the projection $p\colon E \to B$ induce homomorphisms $i_*\colon \pi_n(F, x_0) \to \pi_n(E, x_0)$ and $p_*\colon \pi_n(E, x_0) \to \pi_n(B, y_0)$.

**Theorem 38.** *For any $n > 1$ we can define homomorphisms*

$$\delta\colon \pi_n(B, y_0) \to \pi_{n-1}(F, x_0)$$

*such that the sequence*

$$\cdots \longrightarrow \pi_n(F) \xrightarrow{i_*} \pi_n(E) \xrightarrow{p_*} \pi_n(B) \xrightarrow{\delta} \pi_{n-1}(F) \longrightarrow \cdots \longrightarrow \pi_1(B)$$

*is exact (the base points are omitted for brevity).*

*Proof.* We limit ourselves to defining homomorphisms $\delta$, leaving it to the reader to check that the sequence is exact. Let $\alpha$ be an arbitrary element of $\pi_n(B)$. Then any spheroid $s\colon I^n \to B$ representing it can be viewed as a homotopy $f_t\colon I^{n-1} \to B$, which is endowed with a lift $\tilde{f}_0(x) \equiv x_0$ at the initial moment. Here by $t$ we denote the last coordinate of the Euclidean space $\mathbb{R}^n$ that contains the cube $I^n$. Since the bundle is locally trivial, there exists a lift $\tilde{s}\colon I^n \to F$. Its restriction to the boundary $\partial I^n$ of $I^n$ is an $(n-1)$-dimensional spheroid in $F$, which determines the desired element $\delta(\alpha)$. □

**Exercise 98.** Prove that the above sequence of homotopy groups and their homomorphisms is exact.

**Example.** We know from Exercise 78 that $\pi_1(S^1) = \mathbb{Z}$. Let us prove that all the higher homotopy groups of $S^1$ are trivial. To do this, consider the bundle $p\colon \mathbb{R}^1 \to S^1$ whose projection is given by the formula $p(x) = \exp(x\mathbf{i})$ (here $S^1$ is identified with the set of complex numbers with modulus 1). This is a locally trivial bundle with discrete fibre. This fibre can be denoted by $\mathbb{Z}$. Since $\pi_n(\mathbb{R}^1) = \pi_{n-1}(\mathbb{Z}) = 0$ for all $n > 1$, the portion $\pi_n(\mathbb{R}^1) \xrightarrow{p_*} \pi_n(S^1) \xrightarrow{\delta} \pi_{n-1}(\mathbb{Z})$ of the long exact sequence of the bundle ensures that the groups $\pi_n(S^1)$ are trivial for all $n > 1$.

Recall that the group $\pi_1(S^2)$ is trivial and $\pi_2(S^2) = \mathbb{Z}$ by Theorem 37. Extrapolating from the previous example, it would be natural to think that all the homotopy groups of $S^2$, starting with the third one, are trivial. However, things are more complicated.

**Theorem 39.** $\pi_3(S^2) = \mathbb{Z}$.

*Proof.* The idea of the proof is to construct a locally trivial bundle $p\colon S^3 \to S^2$ and to write the corresponding exact sequence. We construct such a bundle in two ways.

The first method is given as follows. Present the sphere $S^3$ in the form $S^3 = \{(z_1, z_2) \in \mathbb{C}^2\colon |z_1|^2 + |z_2|^2 = 1\}$ and define a free action of the circle $S^1 = \{w \in \mathbb{C}\colon |w| = 1\}$ by the rule $(z_1, z_2) \to (wz_1, wz_2)$. It is easy to check that the quotient space of this action is homeomorphic to the sphere $S^2$ and that the factorization map is a locally trivial bundle.

The second method, which is easier to visualize, is based on presenting $S^3$ as the union of two solid tori with common boundary such that the meridian of one of the tori is identified with the longitude of the other torus, and vice versa. See Theorem 34. Each of these solid tori is fibred into circles (longitudes of type (0,1)). However, these fibrations are not consistent on the common boundary. To get consistent fibrations, we decompose the solid tori into circles of type (1,1), which do not change under the exchange of the meridian and the longitude. The set of all such circles inside each solid torus is parameterized by the points of its meridianal disc. Therefore the quotient space is the union of two discs with a common boundary, *i.e.*, a two-dimensional sphere.

The concluding part of the proof consists in considering the portion $\pi_3(S^1) \to \pi_3(S^3) \to \pi_3(S^2) \to \pi_2(S^1)$ of the exact sequence of the constructed bundle. Since $\pi_3(S^1) = \pi_2(S^1) = 0$, we have $\pi_3(S^2) = \pi_3(S^3) = \mathbb{Z}$ (the latter equality is by Theorem 37). □

## 2.8 Coverings

**Definition.** A bundle $p\colon E \to B$ with connected total space and connected base space is called a *covering* if its fibre is discrete ("discrete" means that the fibre consists of isolated points, *i.e.* each point is an open set).

An example of a covering is shown on page 78. Since a covering is a locally trivial bundle, we can write for it the exact sequence of the homotopy groups. Since the fibre is discrete, all its homotopy groups $\pi_i$ are trivial (except the case $i = 0$, when $\pi_0$ is the set naturally identifiable with the fibre itself).

**Exercise 99.** Prove that if $p\colon E \to B$ is a covering, then the induced homomorphisms $p_*\colon \pi_i(E) \to \pi_i(B)$ are isomorphisms for $i > 1$, and the homomorphism $p_*\colon \pi_1(E) \to \pi_1(B)$ is injective.

Like all locally trivial bundles, coverings possess the homotopy lifting property. It turns out that for coverings this property can be significantly strengthened: a lift of a homotopy is unique (of course, if the lift at the initial moment is fixed).

**Theorem 40.** *Let $p\colon E \to B$ be a covering, and let $\tilde{f}_t, \tilde{f}'_t\colon P \to E$ be lifts of a homotopy $f_t\colon P \to B$ such that $\tilde{f}_t \equiv \tilde{f}'_t$ for $t = 0$. Then $\tilde{f}_t \equiv \tilde{f}'_t$ for all $t$.*

*Proof.* If $P$ consists of just one point $a$, then the uniqueness of a lift is obtained by the usual open-closed argument. Denote by $A$ the set of all points $t \in I$ such that $\tilde{f}(a)_t = \tilde{f}'(a)_t$. This set is closed because the projection $p$ is a continuous map. On the other hand, it is open because any projection is a *local homeomorphism*, *i.e.* it becomes a homeomorphism when restricted to a suitable neighbourhood of each point. Since $A \neq \emptyset$ and the segment is connected, we get $A = I$.

In the case of an arbitrary polyhedron (or even an arbitrary topological space, not necessarily polyhedron) the statement of the theorem is obtained in the obvious way: for any point $a \in P$ the paths $\tilde{f}(a)_t$ and $\tilde{f}'(a)_t$ coincide because they coincide at the initial moment. □

Let $B$ be an arbitrary polyhedron with base point $x_0$. To each covering $p\colon E \to B$ with base point $\tilde{x}_0 \in p^{-1}(x_0)$ we can assign the subgroup $H = p_*(\pi_1(E, \tilde{x}_0))$ of $G = \pi_1(B, x_0)$. As we know, this subgroup is isomorphic to the group $\pi_1(E, \tilde{x}_0)$. Replacing the base point $\tilde{x}_0$ with another point $\tilde{x}_0' \in p^{-1}(x_0)$ yields a subgroup conjugate to $H$.

**Definition.** Two coverings $p\colon E \to B$ and $p'\colon E' \to B$ over the same base are called *equivalent* if there is a homeomorphism $h\colon E \to E'$ such that $p = hp'$ (this condition means that each fibre of the former covering is mapped to the fibre of the latter covering placed over the same point).

**Exercise 100.** Prove that two coverings over the same base are equivalent if and only if the subgroups corresponding to them are conjugate.

This exercise shows that the coverings over a given base are parameterized by the subgroups of the fundamental group of the base considered up to conjugacy. Thus, for full classification of coverings over a given base we lack only the existence theorem. Here it is.

**Theorem 41.** *For each subgroup $H$ of the fundamental group $G$ of a connected polyhedron $B$ there exists a covering $p\colon E \to B$ such that $p_*(\pi_1(E)) = H$.*

*Proof.* Let $K$ be a complex that determines a triangulation of the base space. We assign to each simplex $\sigma \in K$ the set $\{\sigma_\alpha\}$ which consists of copies of this simplex. The copies are parameterized by all possible paths in $B$ that join the base point $x_0 \in B$ with the barycenter of $\sigma$. Two paths $\alpha, \alpha'$ determine the same copy if and only if the loop $\alpha'\alpha^{-1}$ determines an element of $H$. The union of all sets $\{\sigma_\alpha\}$ is denoted by $E'$.

Now we glue the simplices from $E'$ into the total space $E$ according to the following rule: a simplex $\delta_\beta \in E'$ is a face of a simplex $\sigma_\alpha \in E'$ if and only if the simplex $\delta \in K$ is a face of the simplex $\sigma \in K$ and the loop composed of the path $\alpha$, a segment inside $\sigma$ that joins its barycenter with the barycenter of $\delta$, and the path $\beta^{-1}$ determines an element of $H$.

It is easy to check that the projection $p\colon E \to B$ that maps each copy $\sigma_\alpha$ to $\sigma$ is the projection map of a desired covering. □

The conclusion is that *the equivalence classes of coverings over a given base $B$ with fundamental group $G$ are parameterized by the subgroups of $G$ considered up to conjugacy.*

**Exercise 101.** Describe all coverings over the circle.

**Exercise 102.** Suppose that a subgroup $H$ of $G = \pi_1(B, x_0)$ corresponds to a covering $p\colon E \to B$. Prove that the degree of this covering is equal to the index of $H$.

**Definition.** A covering is called *regular* if the corresponding subgroup of the fundamental group of the base space is normal.

There is a simple criterion: a covering is regular if and only if the following property holds: for any loop in the base space, if its lift starting at some point of the fibre is a loop, then any of its other lifts (starting at any other point) is also a loop. Examples of a regular covering and a non-regular covering over the figure-eight are shown in Figure 45.

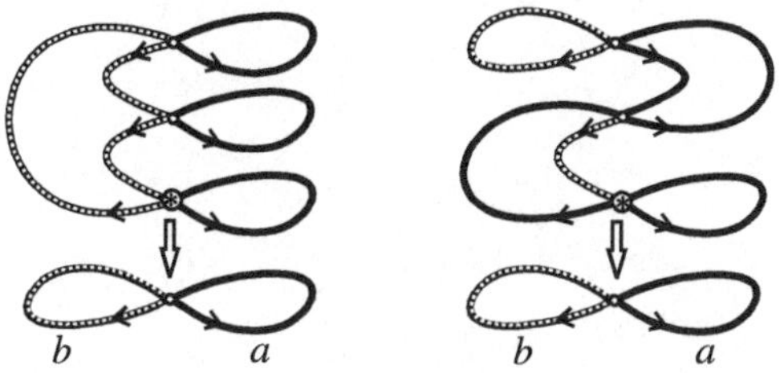


Figure 45. A regular covering (left) and a non-regular covering over the figure-eight.

**Exercise 103.** Suppose that a normal subgroup $H$ of $G = \pi_1(B, x_0)$ corresponds to a regular covering $p\colon E \to B$. Prove that assigning to each oriented loop in $B$ starting at $x_0$ the endpoint of its lift starting at a fixed point $\tilde{x}_0 \in p^{-1}(x_0)$ determines a bijection between the quotient group $G/H$ and the fibre of the covering over the point $x_0$.

To conclude we provide a small informal dictionary between topological and algebraic terms based on the theory of coverings.

| | |
|---|---|
| Space | Group |
| Map | Homomorphism |
| Covering | Subgroup |
| Equivalence of coverings | Conjugacy of subgroups |
| Regular covering | Normal subgroup |
| Degree of a covering | Index of a subgroup |
| Fibre of a regular covering | Quotient group |

Sometimes translating a problem from the topological language to the algebraic one or vice versa makes finding a solution much simpler. For instance, understanding the algebraic proof of the theorem that *any subgroup of a free group is free*, requires a certain effort, while its topological reformulation (*the total space of any covering over a wedge of circles is a one-dimensional cell complex*) is quite evident.

**Exercise 104.** Can the free group with two free generators contain a subgroup of index 3 that is isomorphic to the free group with four free generators?

# Answers, hints, solutions

**Exercise 1** (Solution). Assigning to each group $G$ the set of its elements determines a functor from the category of groups to the category of sets. Functors of this type are called *forgetful functors*. In our example we forget the group structure, *i.e.* the operation in groups. Of course, any forgetful functor is covariant.

A standard example of a covariant functor is obtained by assigning to each (say, finite dimensional) linear space $L$ its dual space $L^*$, *i.e.* the linear space of functionals on $L$. To each linear map $f\colon L_1 \to L_2$ we assign the dual map $f^*\colon L_2^* \to L_1^*$. It is defined by the rule $f^*(\xi)(x) = \xi(f(x))$, where $x \in L_1$ and $\xi \in L_2^*$.

**Exercise 2** (Solution). The groups are certainly distinct, because one of them consists of four elements, and the other one of five elements. What does Theorem 1 have to do with this? Well, that is what we have actually used! The forgetful functor assigns to the groups $\mathbb{Z}_4$, $\mathbb{Z}_5$ different sets (of different cardinalities). Therefore the groups are also distinct.

**Exercise 3** (Hint). We can use the following properties of the determinant: the determinant $|E|$ of the unit matrix is equal to 1 if $|A| > 0$, then $|A^{-1}| > 0$, $|AB| = |A||B|$.

**Exercise 4** (Answer). The rightmost basis.

**Exercise 5** (Hint). Choose two non-equivalent bases and prove that the determinant of the change of coordinates matrix from the third basis to one of them is positive.

**Exercise 6** (Solution). The definition of the linear independence of vectors can be stated as follows: $n$ vectors with the same origin are linearly independent if and only if they are not contained in the same $(n-1)$-dimensional plane. Therefore the independence of points $a_0, a_1, \ldots, a_n$ is equivalent to the linear independence of vectors $\overline{a_0a_1}, \overline{a_0a_2}, \ldots, \overline{a_0a_n}$.

**Exercise 7** (Hint). Use the results of the previous exercise and the properties of the linear independence of vectors.

**Exercise 8** (Hint). Use the results of the previous exercise.

**Exercise 9** (Answer). $C_{n+1}^{m+1} = \frac{(n+1)!}{(m+1)!(n-m)!}$.

**Exercise 10** (Answer). $2^{n+1}$, including the empty face.

**Exercise 11** (Answer). An example of a simplicial complex could be a triangle or the union of two triangles that intersect along one common edge. In both cases the triangle is taken together with all its edges and vertices. Two triangles on the plane that have a common interior point but do not coincide do not form a simplicial complex, even if we add all their edges and vertices.

**Exercise 12** (Answer). $2^7$, since each of the 7 non-empty faces admits two orientations.

**Exercise 13** (Answer). One can take any sequence of groups and homomorphisms which contains two consecutive isomorphisms between non-trivial groups.

**Exercise 14** (Answer). The group of cycles $A_2$ is isomorphic to $\mathbb{Z}$ and is generated by the element $(1, -1)$, the group $A_1$ coincides with the group $C_1 = \mathbb{Z}$, and all the other groups of cycles are zero. All the groups of boundaries are also zero, except the group $B_1$, which is generated by the element 3 of the group $\mathbb{Z}$ and hence isomorphic to $\mathbb{Z}$.

**Exercise 15** (Answer). $H_1(C) = \mathbb{Z}_3$, $H_2(C) = \mathbb{Z}$, and all the other homology groups are trivial.

**Exercise 16** (Answer). $H_m(E(m)) = \mathbb{Z}$, all the other homology groups are trivial.

**Exercise 17** (Answer). $H_m(D(m,k)) = \mathbb{Z}_k$, all the other homology groups are trivial.

**Exercise 18** (Hint). All the groups corresponding to the direct sum of chain complexes are direct sums of the relevant groups corresponding to the individual complexes.

**Exercise 19** (Hint). Use the commutativity of squares.

**Exercise 20** (Hint). The morphisms of the category of chain complexes are chain maps and the morphisms of the category of sequences of Abelian groups are sequences of homomorphisms between respective groups. Of course, to the identity map from a chain complex to itself corresponds the identity morphism of the homology groups, and the property $(\varphi\psi)_* = \varphi_*\psi_*$ holds because it can easily be verified for chains, which represent homology classes.

**Exercise 21** (Hint). Let $K_1$, $K_2$ be two copies of the given complex $K$ endowed with distinct orientations. Assign to each simplex $\sigma \in C_n(K_1)$ the chain $\pm\sigma \in C_n(K_2)$ where the sign is chosen depending on whether the two orientations of $\sigma$ coincide or not. To solve the exercise it is sufficient to prove that this assignment determines an isomorphism between complexes $C(K_1)$ and $C(K_2)$.

**Exercise 22** (Answer and hint). All the homology groups of the segment $I$ and those of the circle $S^1$ are trivial, except the groups $H_0(I) = H_0(S^1) = \mathbb{Z}$ and $H_1(S^1) = \mathbb{Z}$. To solve the exercise we should present both the segment and the circle as simplicial complexes. For instance, the circle can be presented as a triangle (more precisely, its boundary).

**Exercise 23** (Answer). $H_2(S^2) = H_2(T^2) = \mathbb{Z}$, see Theorem 9.

**Exercise 24** (Solution). Since there are no simplices of negative dimension, the group of cycles $A_0(K)$ is a subgroup of the respective chain group. On the other hand, any two vertices joined by an edge yield homologous cycles. Choosing one vertex in each connected component, we get a system of free generators for $H_0$.

**Exercise 25** (Hint). The easiest would be to use the *barycentric coordinates*: each point $x$ of a simplex $\sigma$ with vertices $v_0, \dots, v_n$ can be written in the form $x = \sum_{i=0}^{n} \lambda_i v_i$, where $\lambda_i \geqslant 0$ and $\sum_{i=0}^{n} \lambda_i = 1$. The physical meaning of this expression can be described as follows: if we put masses $\lambda_i$ at the vertices of $\sigma$, then the point $x = \sum_{i=0}^{n} \lambda_i v_i$ is the mass center of the resulting system. A map defined on the vertices can be extended to a map defined on the whole simplex as follows: $f\big(\sum_{i=0}^{n} \lambda_i v_i\big) = \sum_{i=0}^{n} \lambda_i f(v_i)$.

**Exercise 26** (Hint). To construct the extension one could use the barycentric coordinates on simplices, see the hint for Exercise 25.

**Exercise 27** (Hint). A fine point consists in verifying the equality $\varphi\partial = \partial\varphi$ for a simplex $\sigma$ of dimension $n$ that is mapped onto a simplex $\delta$ of dimension $n-1$. In this case exactly two faces of the simplex are mapped onto $\delta$, one preserving the orientation and the other reversing it, see Figure 5.

**Exercise 28** (Solution). The size of a decomposition is measured according to the maximal diameter of its simplices or, equivalently, the maximal length of their edges. The length of an edge is calculated with respect to the standard metric of the ambient Euclidean space $\mathbb{R}^N$. Taking the barycentric subdivision decreases the length of each edge exactly in half. Therefore taking the barycentric subdivision several times, we can make the diameter of the subdivision arbitrarily small.

**Exercise 29** (Solution). Consider the function $s\colon \operatorname{St}(v, K) \to \mathbb{R}$ that assigns to each point $x$ of each simplex $\sigma \subset \operatorname{St}(v, K)$ the barycentric coordinate of $x$ at the vertex $v$. This function is continuous on the closed star. It remains to notice that the open star is determined by the strict inequality $s < 1$.

**Exercise 30** (Solution). If vertices $v_1, \dots, v_k$ span a simplex, then the interior of that simplex is contained in the open star of each vertex. Therefore the stars have non-empty intersection. On the other hand, if the stars do have non-empty intersection then the simplex that contains an interior point of this intersection contains all these vertices.

**Exercise 31** (Hint). Arguing by contradiction, it is easy to construct two sequences of points $x_n, x_n' \in |K|$ such that for any $n$ the points $x_n, x_n'$ are not contained in the same element of the cover, yet $\rho(x_n, x_n') \leqslant 1/n$. By a compactness argument, we can assume that both sequences converge to some point $x_0 \in |K|$, which is contained in some element of the cover, together with all sufficiently close points. This contradicts the choice of sequences $x_n, x_n' \in |K|$.

**Exercise 32** (Hint). The above proof of the absolute version of the simplicial approximation theorem can be repeated without changes. The only additional point which may be needed is the following. If a vertex $v$ of the complex $K_1$ (the subdivided complex $K$) lies in $M$, then the vertex $w = g(v)$ should be contained in $N$, otherwise the inclusion $\mathring{\mathrm{St}}(v, K_1) \subset U_w$ would be impossible.

**Exercise 33** (Solution). For simplicial maps the statements are evident, and in the general case one can apply the simplicial approximation theorem.

**Exercise 34** (Hint). Apply Theorem 1.

**Exercise 35** (Answer). A manifold is called Euclidean, hyperbolic, complex-analytic, conformal, or Lipschitz if it admits an atlas where all changes of charts belong to the respective class of homeomorphisms. In the first case this is the class of isometries of the Euclidean space $\mathbb{R}^3$, in the second, that of isometries of the hyperbolic space $\mathbb{H}^3$, in the third, that of complex-analytic homeomorphisms of $\mathbb{R}^{2k} = \mathbb{C}^k$. In the remaining two cases this is the class of conformal or Lipschitz homeomorphisms (whatever this means).

**Exercise 36** (Solution). The circle is a topological polyhedron, not the actual one. To calculate the degree we replace the circle by an actual polyhedron (with straight sides) in two ways: by the regular inscribed $3n$-gon and by the regular inscribed triangle, see Figure 46. The circle can be identified with each of these polygons via the projections along the radii. Then the simplicial approximation of $f$, viewed as a map from the $3n$-gon to the triangle, acts as follows: on vertices of the $3n$-gon it acts just like $f$, and the sides of the $3n$-gon are linearly mapped to the sides of the triangle. It is obvious that the pre-image of any oriented edge (of any side) of the triangle consists of exactly $n$ coherently oriented sides of the $3n$-gon. Therefore $\deg f = n$.

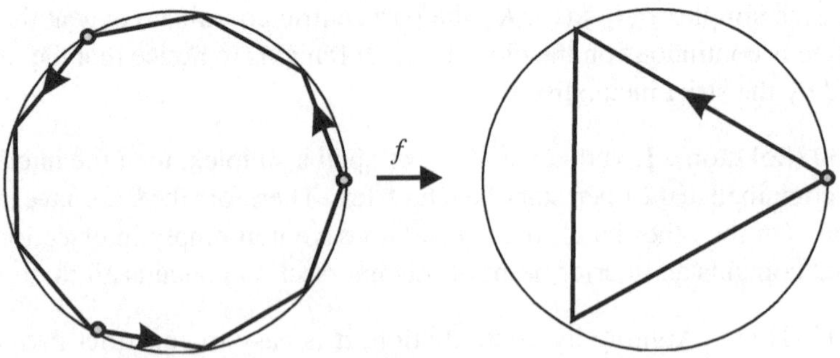


Figure 46. The degree of the 3-fold winding of a circle around a circle is equal to 3.

**Exercise 37** (Solution). The total number of windings coincides with the degree of the map $f : S^1 \to S^1$ that is induced by the central projection of the given path to the outer boundary circle of the annulus. This degree is equal to the difference between the number of the positive pre-images of a regular point and that of the negative ones. If we choose the regular point as it is shown in Figure 47, then these numbers are equal to 3 and 2. Therefore the degree is equal to 1.

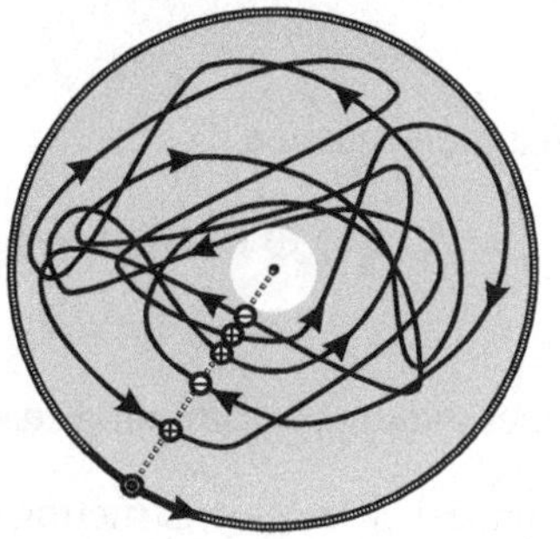

Figure 47. The marked regular point on the outer boundary circle of the annulus has three positive and two negative pre-images.

**Exercise 38** (Answer). Two immersions $f, g\colon S^1 \to S^2$ are regularly homotopic if and only if their writhe numbers $w(f)$, $w(g)$ have the same parity.

**Exercise 39** (Answer). The homomorphism $\varphi$ is a monomorphism, an epimorphism, or an isomorphism, respectively.

**Exercise 40** (Hint). This is a central algebraic theorem on homomorphisms: factorization of a group by the kernel of a homomorphism yields the image of the homomorphism.

**Exercises 41–43** (Hint). All three exercises can be solved by means of diagrammatic search.

**Exercise 44** (Solution). We prove that $H_k(S^n) = 0$ for $k \neq 0, n$ and $H_k(S^n) = \mathbb{Z}$ for $k = 0, n$. To do this we write the exact sequences of the homology groups of the pairs $(S^n, D^n)$ and $(D^n, S^{n-1})$, where $D^n$ is an $n$-dimensional ball embedded in the sphere $S^n$ and $S^{n-1}$ is the sphere of dimension $(n-1)$ (its boundary):

$$\cdots \longrightarrow H_k(D^n) \xrightarrow{(i_k)_*} H_k(S^n) \xrightarrow{(p_k)_*} H_k(S^n, D^n) \xrightarrow{\delta_k} H_{k-1}(D^n) \xrightarrow{(i_{k-1})_*} H_{k-1}(S^n)$$

$$\longrightarrow H_k(D^n) \xrightarrow{(p_k)_*} H_k(D^n, S^{n-1}) \xrightarrow{\delta_k} H_{k-1}(S^{n-1}) \xrightarrow{(i_{k-1})_*} H_{k-1}(D^n) \longrightarrow \cdots$$

Since the ball $D^n$ is homotopy equivalent to the point, we have that $H_i(D^n) = 0$ for $i \neq 0$ and $H_0(D^n) = \mathbb{Z}$. Therefore for $k > 1$ the homomorphism $(p_k)_*$ of the first sequence and the homomorphism $\delta_k$ of the second sequence are isomorphisms. Now we take into account that in both sequences the homomorphisms $(i_0)_*$ have trivial kernels, except the homomorphism $(i_0)_*\colon H_0(S^0) = \mathbb{Z} \oplus \mathbb{Z} \to H_0(D^1) = \mathbb{Z}$, whose kernel is isomorphic to $\mathbb{Z}$. Since the groups $H_k(S^n, D^n)$ and $H_k(D^n, S^{n-1})$ are isomorphic by Theorem 15, it follows that $H_k(S^n) = H_{k-1}(S^{n-1}) = \cdots = H_1(S^{n-k+1}) = H_1(D^{n-k+1}, S^{n-k} = 0$ for all $k \leqslant n$, and the latter group is 0 for $0 < k < n$ and $\mathbb{Z}$ for $0 < k = n$. The equalities $H_k(S^n) = 0$ for $k > n$ are obtained in a similar way: $H_k(S^n) = H_{k-n}(S^0) = 0$.

**Exercise 45** (Hint). Check that the sequence

$$0 \longrightarrow C_n(K_1 \cap K_2) \xrightarrow{i} C_n(K_1) \oplus C_n(K_2) \xrightarrow{j} C_n(K_1 \cup K_2) \longrightarrow 0$$

is exact.

**Exercise 46** (Hint). Use the method of diagrammatic search.

**Exercise 47** (Hint). This is a partial case of constructing a quotient group.

**Exercise 48** (Answer). One row will be added to another one.

**Exercise 49** (Answer). One column will be subtracted from another one.

**Exercise 50** (Answer). All elements of one row change signs, all elements of one column change signs, columns or rows are exchanged.

**Exercise 51** (Answer). $\mathbb{Z}_2 \oplus \mathbb{Z}_8$.

**Exercise 52** (Solution). The elementary transformations of rows and columns do not change the absolute value of the determinant. Hence it suffices to check the statement for diagonal matrices only.

**Exercise 53** (Answer). $H_0 = \mathbb{Z}$, $H_1 = \mathbb{Z}_2 \oplus \mathbb{Z}_2$, $H_2 = 0$, $H_3 = \mathbb{Z}$. All the other homology groups are zero. It is worth noting that the chain complex considered in this exercise corresponds to a cell decomposition of the closed orientable three-dimensional manifold that is obtained by taking the quotient of the sphere $S^3$ by the linear action of the group $\{\pm\mathbf{1}, \pm\mathbf{i}, \pm\mathbf{j}, \pm\mathbf{k}\}$ of the quaternion units. Hence its homology groups coincide with those of that manifold.

**Exercise 54** (Hint). Such a complex can be easily composed from the elementary ones.

**Exercise 55** (Solution). Present the Klein bottle as a cell complex with one vertex, two loops marked by numbers 1, 2, and one two-dimensional cell, whose boundary curve passes along the edges according to the rule $\{1\ 2\ -1\ 2\}$. The resulting relation matrix (2, 0) consists of one row and determines the group $H_1 = \mathbb{Z} \oplus \mathbb{Z}_2$.

**Exercise 56** (Hint). This is a standard fact from linear algebra. The trace is one of the coefficients of the characteristic polynomial of the relevant linear operator and hence does not depend on the choice of the basis.

**Exercise 57** (Solution). Choose a basis in the group

$$\mathrm{Free}(A_1 \oplus A_2) = \mathrm{Free}(A_1) \oplus \mathrm{Free}\, A_2$$

in such a way that elements coming first would form a basis of $\mathrm{Free}(A_1)$, and the remaining one would form a basis of $\mathrm{Free}(A_2)$. Then the upper left corner of the corresponding matrix of $\varphi$ contains a matrix of the endomorphism $\varphi_1$, and the lower right corner contains a matrix of $\varphi_2$. The diagonals of these two corner matrices form the diagonal of the entire matrix, therefore its traces are summed up.

**Exercise 58** (Solution). Evident, since for each simplex there is a corresponding 1 in the diagonal of the matrix of the endomorphism at the relevant dimension.

**Exercise 59** (Solution). See the solution of the previous exercise.

**Exercise 60** (Hint). Apply Theorem 23 and Exercise 58.

**Exercise 61** (Hint). In both cases the homological Lefschetz number is equal to 1, therefore a fixed point always exists.

**Exercise 62** (Answer). $\mathbb{Z}_3$.

**Exercise 63** (Answer). $H_0 = H_1 = H_2 = H_3 = \mathbb{Z}_2$, all the other groups are trivial.

**Exercise 64** (Answer). $\mathbb{Z} \oplus \mathbb{Z}_2, \mathbb{Z}_2 \oplus \mathbb{Z}_2, \mathbb{Z}_3, \mathbb{Q}$.

**Exercise 65** (Answer). $H_0 = H_1 = \mathbb{Z}$, $H_2 = \mathbb{Z}_2$.

**Exercise 66** (Solution). $\partial\partial(a \otimes b) = \partial(\partial a \otimes b + (-1)^{\dim a} a \otimes \partial b) = (-1)^{\dim a - 1} \partial a \otimes \partial b + (-1)^{\dim a} \partial a \otimes \partial b = 0$. In the next-to-last equality we have used the fact that $\partial\partial a = \partial\partial b = 0$.

**Exercise 67** (Hint). First prove the similar equality $(C \otimes D)^* = C^* \otimes D^*$ for arbitrary Abelian groups $C, D$.

**Exercise 68** (Hint). To prove the triviality of $H_2(M)$ one could use isomorphisms $\mathrm{Free}(H_2) = \mathrm{Free}(H^2) = \mathrm{Free}(H_1) = 0$ and $\mathrm{Tor}(H_2) = \mathrm{Tor}(H^3) = \mathrm{Tor}(H_0) = 0$, which are obtained by applying Theorem 27 and the Poincaré duality.

**Exercise 69** (Answer). $\mu \cup \lambda = \pm 1$, depending on the choice of the orientation of the torus.

**Exercise 70** (Answer). 1 and 3.

**Exercise 71** (Solution). If $f_t, g_t \colon [0, 1] \to X$ are homotopies between loops $f_0$ and $f_1$ and loops $g_0$ and $g_1$ then a homotopy between their products can be given by the formula

$$h_t(s) = \begin{cases} f_t(2s), & 0 \leqslant s \leqslant 1/2; \\ g_t(2s-1), & 1/2 \leqslant s \leqslant 1. \end{cases}$$

**Exercise 72** (Hint). Use Figure 34.

**Exercise 73** (Solution). The formula

$$h_t(s) = \begin{cases} f(\frac{2}{t+1}s), & 0 \leqslant s \leqslant \frac{t+1}{2}; \\ s_0, & \frac{t+1}{2} \leqslant s \leqslant 1 \end{cases}$$

defines a homotopy between the loop $f$ and its product by the constant loop.

**Exercise 74** (Hint). Describe a homotopy $h_t\colon [0,1] \to X$, which deforms the product of a given loop and the inverse loop to the constant loop. The loop that is obtained at each intermediate moment $t$ of this deformation behaves as follows: for $0 \leqslant s \leqslant t$ we go along the loop, then stop for awhile, and for $1-t \leqslant s \leqslant 1$ we go along this loop in the opposite direction, *i.e.* go along the inverse loop.

**Exercise 75** (Hint). See the solution of Exercise 71.

**Exercise 76** (Hint). When checking that $\varphi$, $\psi$ are homomorphic and mutually inverse, use the fact that paths $ss^{-1}$ and $s^{-1}s$ are homotopic to the constant path.

**Exercise 77** (Hint). Since the given subset is convex, one can use the same trick as when proving the triviality of the fundamental group of the segment.

**Exercise 78** (Hint). The only difficulty consists in proving that the loop that winds 0 times is homotopic to the constant loop. One of the many possible proofs relies on the simplicial approximation theorem. First we should present the circle as, say, a triangle and approximate the given loop $[0,1] \to S^1$ by a simplicial one. Then we should step by step destroy the situations when a point of the loop passes along a side of the triangle in one direction and then, immediately, in the opposite one. After destroying all such situations, we automatically get the constant loop.

**Exercise 79** (Answer). 16.

**Exercise 80** (Hint). Reflexivity and transitivity are obvious, while symmetry follows from the fact that all moves involved are reversible.

**Exercise 81** (Hint). The first relation implies that the words $ab$ and $ba^{-1}$ are equivalent. Therefore $a^2b = a\,ab = aba^{-1} = ba^{-1}a^{-1} = ba^{-2}$. On the other hand, the words $a^2b$ and $ba^2$ are also equivalent, since each of them is transformed to $b^3$ by one substitution. Therefore the word $ba^{-2}$ is equivalent to the word $ba^2$, which implies that $a^4$ is equivalent to the empty word.

**Exercise 82** (Hint). The associativity of multiplication follows from the fact that appending words is an associative operation, the identity element is the empty word, and the inverse element is obtained by writing the given word from left to right and reversing the signs of all exponents.

**Exercise 83** (Solution). Evident, since the operation of inserting the subword $wR_iw^{-1}$ can be viewed as the consecutive insertion of subwords $ww^{-1}$ and $R_i$.

**Exercise 84** (Solution). Since $x^2 = zx^2z\,zx^2z = zx\,xz^2x\,xz = zxzxz$, we have $x^3 = zxzxzx$. Similarly, $z^2 = xzxzx$ and $z^3 = zxzxzx$, which implies that $x^3 = z^3$.

On the other hand, $x = zx^2z = xz^2x\,x^2z = xz^2x^3z$. Therefore $z^2x^3z = 1$ and $z^3x^3 = 1$. The obtained equalities $x^3 = z^3$ and $z^3x^3 = 1$ yield the desired statement.

**Exercise 85** (Hint). Suppose that presentations of groups $G$ and $G'$ are related by one of the moves. Then an isomorphism between these groups can be given by assigning to each generator of $G$ the corresponding generator of $G'$. The generator $a$ that takes part in moves II, II′, is not taken into account. Of course, one should check that such assignment should transform the relations of one group to correct equalities in the other group, *i.e.* to deducible relations of that group.

**Exercise 86** (Hint). It is easy to check that the assignment $x \to b^{-1}a^{-1}$, $y \to b^{-1}$ transforms the relations $xy^2x = y$, $yx^2y = x$ to true equalities. Therefore this assignment determines a homomorphism from $G$ to the binary tetrahedral group. For checking that the reverse assignment $a \to x^{-1}y$, $b \to y^{-1}$ determines a homomorphism from the binary tetrahedral group to $G$, it is convenient to use Exercise 84.

**Exercise 87** (Hint). Section 1.12 describes how to recognize Abelian groups. In our case not only the groups $\langle a, b \mid a+4b, 2a+3b\rangle$ and $\langle x, y \mid 4y, 3x-2y\rangle$ but also their orders are distinct.

**Exercise 88** (Answer). The free group with two generators.

**Exercise 89** (Answer). $\mathbb{Z}_2$.

**Exercise 90** (Answer). The Klein bottle can be presented as a rectangle $ABCD$ whose sides are glued together according to the rule $AB \to BC$, $CD \to DA$. This decomposition gives rise to the presentation $\langle a, b \mid a^2b^2\rangle$.

**Exercise 91** (Answer). $\langle a_1, b_1, a_2, b_2 \mid a_1b_1a_1^{-1}b_1^{-1}, a_2b_2a_2^{-1}b_2^{-1}\rangle$.

**Exercise 92** (Hint). The space (more precisely, the cell complex) $X$ can be obtained from a wedge of circles (marked by the group generators) by attaching two-dimensional cells as the relations show.

**Exercise 93** (Hint). The associativity of the operation is evident, the identity element is the map taking the entire cube to the base point, and the inverse spheroid is obtained from the given one by taking the composition with the symmetry of the cube with respect to a hyperplane.

**Exercise 94** (Hint). Check that the compositions $\tau_{\alpha^{-1}}\tau_\alpha$ and $\tau_\alpha\tau_{\alpha^{-1}}$ coincide with the identity map.

**Exercise 95** (Hint). The Hurewicz map is well defined because homotopic maps between spaces induce identical homomorphisms of the homology groups. It is homomorphic by the definition of the addition in $\pi_n(X)$.

**Exercise 96** (Hint). Use the fact that any map $f\colon I^n \to X \times Y$ is uniquely determined by the maps $p_1 f\colon I^n \to X$ and $p_2 f\colon I^n \to Y$, where $p_1, p_2$ are the direct sum projections.

**Exercise 97** (Hint). Since the bundle is trivial, all the lifts of the homotopy $f_t\colon P \to B$ are completely determined by maps of the polyhedron $P \times I$ to the base (see the hint for the previous exercise). Therefore to prove the relative homotopy lifting property it is sufficient to use the existence of a retraction $P \times I \to P \cup (Q \times I)$.

**Exercise 98** (Hint). Use the method of diagrammatic search.

**Exercise 99** (Hint). Write down the exact sequence for the covering and use the fact that the groups $\pi_i$ are trivial for $i \geqslant 1$.

**Exercise 100** (Hint). If two coverings are equivalent and the given homeomorphism between their total spaces takes the base point to the base point, then the subgroups corresponding to those coverings coincide. If the subgroups corresponding to two given coverings coincide, then a homeomorphism $h\colon E \to E'$ between their total spaces can be constructed as follows. Each point $x \in E$ is joined by a path $\tilde{s}_t\colon I \to E$ to the base point $x_0 \in E$. Then we lift the projection $s_t = ps_t$ of this path to the space $E'$ obtaining a path $s'_t$ starting at the point $x'_0$. The endpoint of this path is $h(x)$. As was mentioned before, the appearance of conjugate subgroups is caused by having many different possibilities for choice of the base point.

**Exercise 101** (Answer). The $n$-fold windings of a circle around a circle ($n > 0$), and an infinite winding of a line around the circle (see the example of a covering on page 78).

**Exercise 102** (Hint). The definition of $H$ implies that the lift of a loop starting at the point $\tilde{x}_0 \in p^{-1}(x_0)$ ends at the same point if and only if this loop represents an element of $H$. Therefore for any $g \in G$ and $h \in H$ the elements $g$ and $hg$ are represented by the same point of the fibre. This yields a well-defined map $\varphi\colon G/H = \{Hg, g \in G\} \to F$. For the same reason this map is injective. The surjectivity is evident: each point $\tilde{x}' \in F$ corresponds to the projection of the path going from $\tilde{x} \in F$ to $\tilde{x}' \in F$.

**Exercise 103** (Hint). Use the solution of the previous exercise.

**Exercise 104** (Hint). The positive answer is shown in Figure 45. The fundamental group of the total space ("long figure-eight") is the free group with four generators and is contained as a subgroup of index 3 in the fundamental group of the base (the figure-eight), which is also free and has two generators.

# Bibliography

[1] N. M. Bliznyakov, Yu. G. Borisovich, Ya. A. Izrailevich, T. N. Fomenko, *Introduction to differential and algebraic topology*. Kluwer Texts in the Mathematical Sciences 9, Kluwer Academic Publishers Group, Dordrecht 1995.

[2] I. Bucur, A. Deleanu, *Introduction to the theory of categories and functors*. Pure and Applied Mathematics XIX, Interscience Publication John Wiley & Sons, Ltd., London, New York, Sydney 1968.

[3] A. Dold, *Lectures on algebraic topology*. Classics in Mathematics, Springer-Verlag, Berlin 1995.

[4] A. T. Fomenko, D. B. Fuchs, V. L. Gutenmacher, *Homotopic topology*. Translated from the Russian, Akadémiai Kiadó, Budapest 1986.

[5] A. Hatcher, *Algebraic topology*. Cambridge University Press, Cambridge 2002.

[6] P. Hilton, S. Wylie, *Homology theory: an introduction to algebraic topology*. Cambridge University Press, New York 1960.

[7] C. Kosniowski, *A first course in algebraic topology*. Cambridge University Press, Cambridge, New York 1980.

[8] W. Magnus, A. Karras, D. Solitar, *Combinatorial group theory*. Interscience Publishers, New York 1966.

[9] W. Massey, *A basic course in algebraic topology*. Graduate Texts in Mathematics 127, Springer-Verlag, New York 1991.

[10] R. Switzer, *Algebraic topology: homotopy and homology*. Classics in Mathematics, Springer-Verlag, Berlin 2002.

[11] E. Spanier, *Algebraic topology*. 2nd corrected printing of the first edition published by McGraw-Hill in 1966, Springer-Verlag, Berlin 1995.

[12] J. W. Vick, *Homology theory*. An introduction to algebraic topology, 2nd edition, Graduate Texts in Mathematics 145, Springer-Verlag, New York 1994.

# Bibliography

# Index